P9-DOG-459

Good Dog

Also by David DiBenedetto and the Editors of

GARDEN & GUN

The Southerner's Handbook

Good Dog

True Stories of Love, Loss, and Loyalty

DAVID DiBENEDETTO AND THE EDITORS OF
GARDEN&GUN

An Imprint of HarperCollinsPublishers

GOOD DOG. Copyright © 2014 by Garden & Gun, LLC. All rights reserved. Printed in the United States of America. No part of this book may be used or reproduced in any manner whatsoever without written permission except in the case of brief quotations embodied in critical articles and reviews. For information, address HarperCollins Publishers, 195 Broadway, New York, NY 10007.

HarperCollins books may be purchased for educational, business, or sales promotional use. For information, please e-mail the Special Markets Department at SPsales@harpercollins.com.

A hardcover edition of this book was published in 2014 by HarperWave, an imprint of HarperCollins Publishers.

FIRST HARPERWAVE PAPERBACK EDITION PUBLISHED 2015

Designed by Leah Carlson-Stanisic

Illustrations by Clint Hansen

Library of Congress Cataloging-in-Publication Data

The good dog : true stories of love, loss, and loyalty / [edited by] David DiBenedetto & the editors of Garden & Gun.—First edition.

pages cm

ISBN: 978-0-06-224235-8

1. Dogs—Anecdotes. 2. Human-animal relationships—Anecdotes. I. DiBenedetto, David. II. Garden & gun.

SF426.2.G663 2014

636.7—dc23 2014015874

ISBN: 978-0-06-224237-2 (pbk.)

15 16 17 18 19 OV/RRD 10 9 8 7 6 5 4 3 2 1

If a dog will not come to you after having looked you in the face, you should go home and examine your conscience.

—Woodrow Wilson

Contents

Introduction:

My Life in Dog Years

As I write this, my five-month-old son, Sam, is lying on a blanket in the living room, surrounded by toys. He's cooing and babbling and kicking his legs. Of course, I'm watching his antics with delight, marveling at the changes each day brings to his coordination and personality. But my eyes are not the only ones on Sam. Just off the edge of the blanket my Boykin spaniel, Pritchard, lies with her head on her paws, tuned to the action.

Like most dads, I have big dreams for my son. Sam's youth will be filled with love, encouragement, and exploration. He will also never be without a dog. I believe nothing builds a kid's confidence and sense of worth like caring for a canine.

Thankfully my parents weren't cat people. The first dog of my childhood was a mutt named Flap Jack. He came by way of a family friend who lived well beyond the city limits of our Savannah, Georgia, home. The product of a tryst between local strays, Flap Jack was flea ridden and frightened of a human shadow when we picked him up. My oldest brother, Bob, who had valiantly lobbied my parents for a dog, couldn't get close enough to pet Flap Jack for a week. And though my parents had given in to Bob's pleas for a pup, there was no chance the animal would be allowed in the house. I watched as Bob, with the help of our next-door neighbor, built a run and a doghouse, complete with a shingled roof. And I watched as the frightened little pup eventually learned his name, found trust in those who fed him, and became a dog. I followed the two of them around as little brothers do. And I remember the day, some years later, when Bob told me he was going to college and that Flap Jack would now be my dog, my responsibility. I recall exactly where I was standing in the yard and the sound of the cicadas and the feeling that I had become something altogether different than I was.

A few years later, when I graduated from eighth grade, my parents gave me a yellow Labrador retriever I named Salty Dog. I had dreams of running Salty in field trials and wading into duck swamps with him in the predawn darkness. But when Salty was just over a year old, he developed a dangerous habit. He would often bolt from the yard in the middle of a training session and not return for hours, sometimes days. Our vet suggested that neutering Salty would solve the problem. But a few weeks after the big snip, I was tossing a ball for Salty when he lit out for parts unknown. I went running after him, but my legs were no match. As usual, my mom and I piled in the minivan and went looking for him. We had no luck. Back at home, there was a message from the vet on the answering machine. Turns out Salty had shown up outside his office door–a two-mile trip that involved traversing a busy road–slobbering on the glass until they invited him in.

We picked Salty up and returned home. A few days later, he hightailed it yet again, and about an hour later the vet called. A couple of weeks after that, the same scenario played itself out. This time my dad drove me to the vet. On the way home, I asked my dad why a dog would be so intent on running to the vet's office when most dogs hated going. With a bit of hesitation, my old man answered, "Maybe he's looking for his balls."

Salty never lived up to my field trial hopes, but he did live up to his name, accompanying me throughout my youth in a jon boat on the coastal waters. He died many years later after I had graduated from college and moved to New York City to become a magazine editor. Even though his Houdini-like escapes had long since stopped (initially curbed by an electric fence and later old age), I had decided that an NYC apartment was not the place for a dog used to napping in the warm sun and tormenting squirrels at his leisure. The teary phone call from my mother not only left me crying but also dogless for the first time since I was a child.

That void would not be filled until the arrival of Pritchard. I had left NYC for a position at *Garden & Gun* in Charleston, South Carolina. My bride to be, whose roots ran centuries deep in New Canaan, Connecticut, was unsure about the move, but her spirits were buoyed by the promise of a dog (and warm winters). We quickly settled on a breed, and not long after we had six pounds of wriggling puppy. Pritch was the first being that we loved together. And we spoiled her rotten. I once again dreamed of training a world-class gundog, but Pritch had other ideas (mainly sneaking onto the bed for a nap when no one was watching). Still, she loves to be in the field. And one day soon, if I'm lucky, Sam and I will head outdoors together, following Pritch into the woods.

Around the *Garden & Gun* offices we often joke that the magazine's Holy Trinity is bourbon, dogs, and barbecue, but dogs truly reign supreme. Since the magazine's first issue (Spring 2007), we've included a column called Good Dog, and from the start it's been an overwhelming reader favorite. The concept is simple: Find great writers who want to tell stories about their dogs. The canines can be purebreds or mutts, good or bad, living or dead. And some of the best writers in the country have answered the call. I challenge you to get through John Ed Bradley's tale of his smelly, drooling bulldog without shedding a tear. Or finish Bronwen Dickey's paean to pit bulls without changing your tune on the breed. Or make it through just a few paragraphs of Jonathan Miles's essay about his failed bird dog without laughing so loud you'll make sure no one is looking. For this book we compiled the best of the best and added a bunch of great new essays too.

I'm hoping my boy, like most of the writers in this book, will grow up to be a dog person. And if he calls me one day to ask whether I think he should get a dog of his own (or one for his kids), I'll tell him what my friend Guy Martin wrote in these pages: "It's simultaneously *never* the right time for a new dog, no matter what, and *always* the right time for a new dog, no matter what." In other words, get the dog. I hope he listens.

David DiBenedetto
Editor in Chief
Garden & Gun
Charleston, South Carolina

CHAPTER 1

The Troublemakers

Appetite for Destruction

BY T. EDWARD NICKENS

She marked the beginning of our marriage, arriving just a few months after we walked down the aisle, during that heady time when life seems so full of hope and possibility. But no way were we having a kid, not yet, so Sweet Emma Pearl curled up in a little yellow ball and rode, whimpering, in Julie's lap the entire four-hour drive home from my buddy's house. Seven weeks old. She slept in our bed and lay down beside the tub whenever either of us took a shower and laid her head in Julie's lap for every meal we ever had at home for close to a decade. "It's not dirt if it came off Emma's paws," Julie liked to say, even when the dirt was swamp muck or the red clay from a Piedmont dove field. We loved that dog as only a young couple starting out in the world can love a dog that's all their own. So to us it was only a charming part of her canine nature that she had the appetite of a goat.

We've all heard about hounds that can't stay out of the garbage or refrain from chewing the television remote to bits. But Emma's gustatory excesses went way beyond a lack of self-control. She was hardly a year old when we realized there was more to her munching than a puppy's normal and expected oral fixations. Her first serious meal of something that only marginally resembled food was a three-pound bag of self-rising flour. We came home to find her on the kitchen floor, panting, face dusted as white as a mime's, muzzle caked with marble-size globs of Martha White's best. She was panting from thirst, and we fed her teaspoons of water at a time, fretting about the giant loaf of bread rising in her belly.

And it was on. Over her nine years, no amount of yelling, pleading, or chasing could keep her from ingesting whatever struck her fancy. She would cower at my thunderous ovations of rage, licking her lips. I doused the trash can with cayenne pepper, which she enjoyed immensely. I mined the kitchen counters with mousetraps to snap some sense into her paws. Nothing doing.

She ate onions beyond counting. She would hide them under her dog bed and gum them in the middle of the night.

Used Q-tips were a treasure not to be rushed, but slowly gnawed on like cud.

She could pick individual blackberries with her lips.

Once, I took a frozen smoked turkey breast out of the freezer, set it on the counter, then ran to the grocery store. When I returned, less than twenty minutes later, there was no turkey breast to be found. Just one tiny corner of a plastic bag and a single sprig of rosemary on the floor. Emma looked as if she'd been blown up like a pool toy. In the time it took me to buy a box of rice, she'd eaten close to eight pounds of turkey breast, bone, and plastic bag. All of it frozen hard as granite. She didn't even break a tooth.

At the time, this was all a funny sideshow, but as Emma's tastes were emboldened–and once we had kids–her feats of ingestion became more problematic. Our children learned to walk around the house like gibbons, arms stretched overhead, fingers grasping crackers out of Emma's reach. Our solution to dirty diaper storage would have passed the security protocols at Fort Knox. She had a love for beaver poop that was undeniable. This was a particular offense to my hunting buddy, who often shared a seat with Emma on the odiferous ride home from the duck swamp.

Perhaps her most impressive infraction occurred the year I started fly fishing for striped bass. On my Christmas wish list were dyed bucktails, flashy tinsels, and strips of glittery foil–just a few of the ingredients for Clouser Minnows. "It never occurred to me," my wife has said, many times since, "that the dog would be the least bit interested."

When Julie walked into the house, she thought she'd stumbled onto a murder scene. In the middle of the living room, Emma had feasted on a five-pack of bright red bucktails, grinding the crimson ink into a three-foot circle of carpet. Elsewhere, blue and green orbs of color marred the floor, and everywhere was the shrapnel of chewed-up tinsel and foil. We had to rip up the Berber carpet. For a week I shoveled brilliantly ornamented poop from the backyard.

I'm absolutely convinced that Emma's garbage-hound habits led to her ultimate demise. The official diagnosis was pancreatitis, and we never wondered what might have caused *that*. The first time it hit her hard, I was in the Alaska backcountry, unreachable for two weeks. Julie rushed Emma to the veterinarian with a request that he was all too willing to grant. "I don't care how much it costs," Julie told him. "But keep this dog alive until my husband gets home."

She held on for another week and a half, then collapsed on the living room floor. This time Emma curled up in a big yellow ball, and lay whimpering, once again, in Julie's lap. We sat there together, the three of us on the floor, just like in the old days, marveling at all the love we'd managed to hold since Sweet Emma Pearl came home.

The next morning I had to cook seven pigs for a church barbecue, and once I got the coals started I carried Emma to the car, and Julie and I drove her to the vet. I dug Sweet Emma's grave while bawling like a baby, hands slicked with hog fat, the vinegary tang of barbecue sauce wafting over the backyard.

Every now and then I catch a glimpse of her headstone in the backyard under the tulip poplar, and I get the urge to take a greasy paper towel or the cut-off ends of onions and toss them on the grave. Some might figure that for a sacrilege. But I'm pretty sure my Emma would appreciate the gesture.

Sadie, the White Devil

BY ALLISON GLOCK

I grew up with outdoor dogs. This was the way it was back then, where I was from. Dogs were beloved, but not welcome on the couch or in the kitchen. The dogs of my youth were thus matted, sandy, and often drooling. Dogs with mud caked in their fur and flea-bitten noses. Dogs that rode untethered in the backs of flatbed trucks. Dogs that wouldn't know a canine sweater if they swallowed one. Dogs that, when injured, were not taken to the vet for $3,300 hip-replacement surgeries, but left to *adapt* or, if they couldn't, walked one last time out behind the garden shed.

My mother's West Virginia family had no dogs, as even a free dog costs money. So when I was allowed to adopt my first puppy—a mutt I named Taffy in an unusual fit of girlishness—I also got a long lecture about finances and responsibilities. Taffy was more than just my first dog. She was an *extravagance*. Even so, Taffy was mostly relegated to the yard. As were all the dogs that followed. My parents were not what have come to be known as "dog people."

"You don't want to deal with the havoc of an indoor dog!" my mother would advise. "The mess alone."

It wasn't until I graduated from college that I adopted my first real dog, which is to say, the first animal I would truly raise on my own. Indoors. The way real dog people do.

I was living in Knoxville, Tennessee, and a friend told me about a shelter just outside of town that specialized in "hard to place" animals. Why I believed I was equipped for a "hard to place" animal having had essentially no experience with pets beyond refilling the outside water bucket and picking poop out of the yard can only be chalked up to the willful arrogance of youth.

On my initial visit to the shelter, I met several adoption candidates. The volunteers, who were not unaware of my limitations, steered me toward "mature" dogs that seemed really, really tired.

"Buttercup is a sweet old girl who is happiest in her bed," one said as I stroked Buttercup's wiry chin. "And we also have Jake," she continued. "He is almost blind, really gentle, very passive."

I liked Buttercup and Jake. I did. But while I was weighing the pros and cons of a blind dog versus an inert dog, another candidate announced herself. She did this by lunging at my thigh.

The volunteers gasped, quickly pulling her off, breathlessly asking how she got out and ushering her back to her "special place" far from other dogs and visiting strangers.

"What's her story?" I asked.

The workers shot each other panicked glances.

"Oh, Pandy is not for you. She has some unique challenges."

"Like what? Her name?" I joked, eyeing Pandy, who was vibrating with hostility in her pen, staring at me as if I were raw hamburger in a people suit.

"She was badly abused," a volunteer said. "And it has left a lot of emotional scarring. She will be a handful for even the most skilled dog trainer."

And there it was. The gauntlet.

"Can I make an appointment to see her again?"

The volunteers conferred in the back of the room like car salesmen. Eventually, the head shelter worker emerged with a ream of forms.

"Fill these out, then we'll see."

Apparently adopting a potentially life-threatening dog required additional paperwork. I went home and dutifully penned in every line. If they thought they were going to put me off adopting Cujo 2.0, they were sadly mistaken. I was stubborn. And strangely drawn to Pandy. She was beautiful, as many crazy females are. Her coat was white with black patches, her nose long, proud. She was an Australian shepherd mix, a narrow, tall dog, nearly my five-ten height when on her hind legs. Her eyes sat wide on her face, making her whole mien shark-like. She moved in the same spirit, determined and fierce and breathtaking to observe. It didn't deter me that she had snapped at me on first sight. Many of history's most enduring love relationships began just that way. Besides, I knew no one else would ever adopt her. And the shelter folks knew it too.

I took Pandy home a few weeks later, after a labored approval process and several more visits, some more harrowing than others. Once home, Pandy became Sadie, a fresh start for her and for me.

I'd like to say that after she was settled and nurtured and

taken to many, many obedience and canine behavior modification classes, Sadie chilled out and became the sweet companion I craved. She did not. Sadie stayed insane. Just as with humans, some damage cannot be undone, only managed. So I adjusted. I quickly learned to handle her quirks and complexities, to jump through countless hoops to avoid causing her even a whiff of distress, lest she explode. At times, I felt like I was dating Naomi Campbell.

We avoided things, like backpacks, and tinfoil, and loud noises, and other dogs, and bright light, and Christmas trees, and sticks, and wet grass, and my mother, and squirrels, and my roommates when they'd been drinking, which was pretty much all the time.

Sadie was a deeply cynical dog. She wouldn't accept food from a stranger. She didn't like unfamiliar people touching her. And she was very clear about her boundaries. You could be stroking her back without issue for two minutes, but once Sadie decided it was enough, you'd better stop or risk losing your ring finger.

Regrettably, there was no way to know every time Sadie was ready to freak. With mental illness comes a certain level of built-in spontaneity. One afternoon she charged a potential suitor because he was wearing a baseball cap. Sadie had met the boy before, but the cap triggered her madness, and in an instant, she went after him, teeth bared, an attack that sent all six-foot-three, 280 pounds of him tumbling into the protective cover of a nearby hedge.

"It was the hat!" I said, running after him. "She hates hats!"

I apologized, but the damage was done. My suitor (and his hats) stopped coming around.

Which was probably for the best, because Sadie had other troubling habits. Like nervous pacing around the perimeter of the room. And growling outside the bathroom door whenever anybody besides me was inside.

Sadie never bit anyone, but it wasn't for lack of trying. So legendary were Sadie's freak-outs, my friends dubbed her the White Devil. Describing her as "high maintenance" would be like calling Bravo's housewives of New Jersey "repugnant"—a description woefully short of the mark.

And yet.

I loved that dog more than any dog I have shared my time with before or since. I had to earn her affection, but once I did, Sadie was loyal to a fault. She was protective. And smart. Sadie appreciated my efforts to give her a better life and never forgot the abuse she'd survived before. More simian than canine, she was like a research chimp clever enough to realize the crappy hand she'd been dealt. No pushover, Sadie knew the score. Life is hard. People can't be trusted. Vigilance is key. Be wise about whom you love, and when you do love, do it with every fiber of your being. Till death do you part.

And so it was with Sadie and me.

A few months after my second child was born, Sadie became ill. We'd been together for more than ten years, and she was suffering greatly. In our last photograph, taken hours before she died, I am on the floor, wrapped around her like a blanket, my one-year-old reaching her tiny hand toward Sadie's barely open mouth. (Sadie was never aggressive toward children. She knew, I think, where goodness lived.)

When the vet told me I had to put Sadie down, something inside me broke. Well-meaning friends suggested maybe I should be "relieved." She was, after all, a lot of work. Which was true. But I was not relieved.

I have had more dogs since Sadie. Right now I live with two rescue pit bulls who are joyful and goofy. They chase balls and frolic and never snarl at anyone. They are normal, good-natured dogs who don't cause me a moment's worry. And I love them too. But it isn't the same. My pits will never break my heart. That distinction is reserved for Sadie. The White Devil. The first dog I dared to let inside.

The Canine Criminal

BY C. J. CHIVERS

When some people remember the dogs in their past, they remember play and affection and hunting trips. Me? I remember a felon. Max, we called him. He started out with promise but in a few swift years evolved into a lean, one-eyed, battle-scarred epileptic with a record of time served and a list of enemies earned.

Time erodes memory, and certain details of Max's crime spree are lost to the decades. When exactly was he born? No one can say. It was the mid-1970s, and our parents had decided that our family was due for a dog. So we drove one day in the station wagon to a veterinarian's office outside of Binghamton, New York, where we lived. There the vet—we called him Doctor—ran a small kennel. The doctor sold puppies cheap: ten dollars a head.

In a stinking room, boxed off with fencing into separate cells, were scrums of small, yipping dogs. Among them was Max—a skinny mixed-breed hound, mostly black, with a whippy tail, a white neck, and brown spots above his dark eyes. He had an underbite on one side, which gave him a world-wise air, as if he had already done time. And he could leap–higher than the other dogs competing for attention. We selected him immediately, struck by his electric energy. He was also small–destined not to top thirty pounds, but rippling with muscle, even as a pup. He seemed the canine equivalent of a lightweight boxer. This was the first of many deceptions. Max had not, we would learn, acquired the build of a disciplined prizefighter. He had the rough-hewn frame of the inmate, the reprobate training for escape.

We brought him home.

My father had been raised with hunting dogs, and later, because his father had been blinded in a hunting accident, he had lived with Seeing Eye dogs, too. He expected rules. Rules were imposed. Max could not venture upstairs. He could not put a paw on furniture. He was quickly house-trained. Binghamton had a leash law, and a dogcatcher, and so Max was expected to stay home, fenced in the yard except when out with the kids, when he was to be kept close.

Like many rules, all of these sounded sensible on day one. But a home is no Alcatraz. Any notion of confining Max faced a reality: Max. Our yard was fenced. Max dug under the fence and roamed. We retrieved him each time, sometimes after hours, other times after days. A new plan was found. We would drive a stake into the soil and tie him to it, with perhaps twenty-five or thirty feet of play. Max briefly puzzled through this new form of confinement, wearing the grass into mud. Then he forced his will upon his circumstances and bent the world back to his command. First he pulled up the stakes, even corkscrew-shaped stakes twisted into the soil. Again and again he ran free, dragging a rope behind him. We kept trying. So did Max. One day, he ended any notion of confinement by chain or rope. After tying him to a post, we returned home to find no dog. Instead, the rope was leading up over the six-foot stockade fence. This was a chilling moment for a boy aged ten or twelve. My dog, it seemed, had hung himself. He was a prison suicide. But when we scrambled to the property line and looked down into our neighbor's yard, we found only his collar hanging there, in the air, several feet above the ground. Max had slipped through and dropped to the opposite side–gone.

If the yard could not contain Max, neither could the house. To each of us, Max was a delight–a deeply affectionate friend, a rascal with charm. But he was his own dog, and if he drew physical and social sustenance from us, he communed as well in another world. He was, and let's be clear, an animal more comfortable in the animal realm. We, it seemed, were the bandit's safe house. He would run himself hard outside and come inside for food and sleep. Sleep was sometimes one of his many deceptions. When Max was ready to go outside, he was ready to go outside. When we

opened the door–say, to check for mail, or to head to school–he would dash past our shins with an athleticism and determination that bordered upon maniacal. Sometimes we managed to catch him as he exploded past. But usually our hands slipped off of him as if he were a greased eel.

Once outside, Max was the real Max. It was not pretty, and those who lived near us must have thought us mad. He raided our neighbors' garbage, knocking over cans and scattering scraps. He hunted anything smaller than he was. He leaped into the air to snatch flies with a snap of his snout. Once we saw him attack a nest of yellow jackets. (Let's just say he was enthusiastic, clever, and addicted to an adrenaline buzz. None of this should be confused with smarts.) He strutted past the fences that confined the neighborhood's other dogs, and taunted them. But eventually Max would grow bored and walk away, the unconfined prince of Johnson Avenue. *What are you going to do about it?* This could not end happily, and it didn't. One day he staggered home one-eyed, having been chased off a lawn by a neighbor swinging a rake. The doctor sewed him up, and soon epilepsy set in, apparently from the head trauma. Nighttime seizures became an occasional routine; we moved aside the furniture while he twisted and rolled. After a few minutes, the seizures would pass. Max would be Max again, trying to get out.

Neither food nor hunting was enough for him. Epilepsy and limited vision did not dampen his wilder impulses. Besting bigger dogs was only a game. A fuller account must be honest: Max wanted sex. For it he prowled the entire town. Sometimes we would spot him miles from home, moving swiftly, with a drive only he could explain. We would retrieve him, and he would try again his domestic life. But he was a soul on parole, and his presence in our house never lasted. Always he would bolt. One neighborhood dog, Moxie, was almost endlessly in heat, and Max cycled through her affections each spring and fall. The city dogcatcher knew Moxie, too, and he regularly trolled through the crowd of male dogs that assembled where she lived. These were usually the times Max ended up in the pound. We had a rough sense of the timing. If Max was gone two or three days, no need to worry. After four or more days, we could visit the pound and find him there, covered in fleas, contemptuous in his cell with his underbite and one-eyed leer, ready for my parents to pay the fine and forgive.

As Max broke down the rules, we grew lax. What was the point, exactly, of trying to chain a dog in the yard if he simply climbed the fence and jumped into the air? And how could you contain a dog that pretended to sleep, biding time for a crack in the door? The answer was: We couldn't. We didn't. We failed.

And sometimes our math was wrong. One time Max vanished for several days, and he was not at the pound. Moxie was in heat a half block away, and the usual crowd milled on her lawn. We were stumped. Max would not have headed to another side of town with his date waiting nearby. So where was he? Several walks past the neighbor's house turned up no clue. Then came the breakthrough. While walking past Moxie's home and once more scanning the crowd of desperate-looking dogs for mine, I heard Max. His bark was distinctive. But he seemed to have made himself invisible. He could be heard but not seen. I called for him. He answered back. I called again. He answered again. I ventured onto the lawn, nearer to my neighbor's home, and at last the latest in Max's misadventures was revealed. A basement window had been broken and a piece of plywood fitted over the open pane. This had been pushed aside, allowing a shaft of light to fall into the gloom. I poked my head near and looked down. There, looking up, was Max, giving me his one-eyed, where-have-you-been stare. Moxie, his exhausted concubine, was lying on her side beside him, panting. A few minutes later, I was explaining to the neighbors that my dog had plunged through their basement window and was in residence in their cellar, impregnating Moxie, whom they had responsibly kept inside. Max came home for another brief stay. He needed, if nothing else, to catch up on sleep.

Hurricane Muffin

BY KATIE CROUCH

He wasn't a good dog. That was the strangest part. Even when he was a puppy, Muffin, our cairn terrier, was yappy and mean, calculating and chewy.

His unpleasant personality may, in fairness, not have been his fault. Our block was a particularly staid stretch of South of Broad Charleston, South Carolina—the sort of place where a bench painted the wrong shade of green could cause a neighborhood uproar. An old-money street, frigid toward children, where shining hunting spaniels shadowed their masters with purpose. These dogs bore elegant, worthy names like Plantagenet, Woodrow, and Artemis. How could a dog, a terrier, called Muffin exist in such a place without holding a grudge?

My parents had moved us from the North (we quickly learned to capitalize the region, as if it were a different country) for university and medical jobs. Mom and Dad were geeks; they didn't care about fitting in. They bought our rambling Victorian on the cheap from an old Charleston family who were unloading the place after a Faulkneresque bout of ruin and suicide. "Just ignore the bloodstains on the carpet," my father liked to say to guests, secretly testing how long it took them to leave after that statement sank in.

Even among the new rules of etiquette, my parents still clung to their New York senses of humor. And when it came to populating their Southern manor, they wanted a *funny* dog. It wasn't that they didn't like well-behaved, reasonable canines. But why, they said, pretend to be something you're not? After all, we were terrier people. We shrieked, we barked, we scratched places better left untouched.

It was my father's job, one Christmas Eve in the 1970s, to procure the specimen from a litter "out in the country." *Highway 17 south to the peanut stand,* the directions read. *Turn at the dirt road, and honk when you see the blue trailer. If you get lost, DO NOT get out of your car. Bring dimes. Go back to the gas station in Walterboro and call.*

The bundle seemed correct–fluffy, waggy, all that. My father felt good about handing over the two hundred dollars. There were papers, and the breeders looked, if not friendly, exactly, dog literate. Then, just as the pair hit the Ashley River bridge, the new dog leaped through the air and attached his adorable puppy teeth–all twenty-eight of them–an inch deep into my father's exposed wrist.

You see, no matter how much we tried to ignore it, the thing was that Muffin sort of sucked. "This one's a nipper!" my mother cried on Christmas morning as my brother's toes bled through his *Star Wars* slippers. "It's nothing. *All* puppies do that. Just don't put him near your eyes, nose, ears, or any other parts you happen to like."

Oh, Muffin. Bred to stalk rodents in the cool hills of Scotland, the dog lost his inner quest to move in the Southern climate. By three, his belly swelled over his legs, giving him the look of a hairy manatee. Despite thousands spent on special flea medicine, he developed a skin condition that caused incessant itching, quelled only by humping the furniture. A stench resembling that of a bloated whale persisted despite weekly shampoos. Watchdog ability? After sleeping through the day and evening, he would spring alive at 3:00 a.m., announcing, with a piercing yap, every gnat that happened by the back door screen.

It wasn't that we didn't love the animal. In those early years, my brother and I were outsiders. After long days of being shunned at school, we would limp home and throw our arms around our greasy canine life raft. But Muffin was indifferent to our affection. In fact, as if to prove this, every month or so he would rouse himself enough to shoot out the front door. My father, who by now was eyeing the Woodrows of the block with envy, would quietly shrug his shoulders and let him leave. Yet Muffin always returned, most likely due to the collar my mother had gotten made: *If found, please muzzle and bring back to 50 Church Street.*

On the day after my sixteenth birthday–September 21, 1989–Hurricane Hugo pressed down on what was, by then, our city. My mother, ever the scientist, insisted that the data showed it would not hit Charleston directly. Those people leaving were fools! She filled the bathtub with water and had my father close the shutters. We didn't know that staying was a terrible idea until the gathered news trucks took off. “We’re out of here,” the reporters shouted to my brother and me over the rising wind. “Hope we see you on the other side!”

As the rain started in, my mother allowed us to drink beer. She sent us upstairs with one can of Coors each, thinking the alcohol would put us to sleep. What she didn’t realize was that there is no way to sleep through a hurricane because there is a locomotive of wind hitting on four sides. For an hour or so my brother and I sat in our beds on the top floor, watching our bedroom walls dance in and out like Jell-O. These were lovely rooms, usually full of light due to the large, plentiful windows. Now the rain was coming down so fast it looked as if a fire hose were pointed toward the glass.

And then, Muffin woke out of his coma and–for the first and only time in his life–did something right. He ran upstairs to our rooms. He barked. He circled. He told us, in no uncertain terms, to get the hell away from the windows and to come with him, to the dark, airless hallway in the bottom center of the house.

“Follow Muffin!” my father cried. It seemed a slim hope, to trail a stinky terrier with no credentials, but we did it, bringing our pillows and blankets to the spot on the first floor where he commanded us to settle. Minutes later, windows began to pop like balloons in the bedrooms we’d previously occupied. In the back of the house, where earlier my brother had been listening to the weather radio, the roof blew off with a sickening crack. The water rose through the basement, licking the front door, yet stopped short of the landing the dog had chosen for us.

Did Muffin actually save us? Probably not. But he did get us moving and kept us away from the glass. If we ventured away from him, he growled. When my father peeked out the door during the eye of the storm, he barked. It was as if, for ten hours, he were taken over by the soul of Old Yeller. He actually *gave* a crap.

We sat huddled as a clan, foolish outliers, our hopes pinned to an animal no one believed in. Then the wind died, the sun came out, and the town stank of pluff mud. We were together, alive.

At this point, Muffin waddled back into his career of sucking. He pissed on carpets and ran away until, at nineteen, he gave one last bitter sigh and died. Now, years later, we have all spun off into other orbits. The old house is sold, though my mother lives around the corner in a more modern version. She has cairn terriers again, two of them. They gnaw on doors and have to be kept away from her grandchildren because they bite babies. "Careful," she said when we met, "this one nips."

Escape Artist

BY BOB McDILL

While I was growing up, my family owned a succession of dogs. Some we acquired on purpose and some by accident. Among the purebreds were a red chow, a beagle, a collie, and a big bluetick coonhound. I don't know why my dad wanted to own a bluetick coonhound. That he also owned a one-eyed saddle horse may somehow put it in context. Blueticks get their name from the blue-black spots or "ticking" on their coats. My older brother and I named him Spot.

Spot turned out to be of the extra-large variety of bluetick. When he reached maturity, or doghood, he set about to whip every other dog in the area. This he accomplished. But he whipped only the males, of course. Concerning the females he was reported to be a rake. Picture a hound dog with a smoking jacket and a snifter of brandy.

Dogs are important to little boys. For many grown-ups a dog is an instrument for finding a covey of quail in a Georgia thicket or retrieving a duck in an icy marsh or pooping on an antique rug. But for a little boy a dog is a companion, a pal. I have an old black-and-white photo somewhere of Spot and me "dancing." He's on his hind legs and I'm holding his front paws in a perfect waltz square as we swirl around the lawn. He's a head taller than I. He may have been leading.

Spot, like certain other dogs I've heard about, had a very accurate inner clock. According to my mother's often told tale, every afternoon exactly five minutes before the school bus was scheduled to arrive, he would get up from his bed in the garage and walk out to the road. When my brother and I stepped down off the bus, Spot was there to greet us. And on his face was what appeared to be a wide grin.

In those days deer hunting with dogs was still legal in parts of Texas. Daddy occasionally hunted in Texas's Big Thicket with an acquaintance of his whom he described as "a very rich and prominent friend." On one particular hunt my father added Spot to the man's pack of hounds. According to Daddy, Spot put all the other dogs to shame. He reportedly chased a stag from Village A all the way to Village Z, a distance of several light-years. The "rich and prominent friend" then offered Daddy a vast fortune for Spot. My father of course turned down this great sum, explaining that he could never accept money in exchange for "a member of the family." But my father was a teller of tales, most of which made him appear to be a grand gentleman of great courage and munificent kindness.

Spot ate a tremendous amount of food, but only if it was food he liked.

I won't call him a picky eater, but he had a preference for expensive dog food, the kind that comes in a container rather than a bag. When presented with dry dog food, he'd nibble a few pieces, then look up at you with a puzzled expression as if to say, "These are fine munchies for cocktail hour, but where's dinner?" His weight would fall off. Finally we found a solution: gravy. Spot preferred chicken gravy on his dry dog food, but any kind of gravy would do. Mother was a fine Southern cook. We had plenty of gravy.

But during those years our little country hamlet was morphing into a suburb. The old properties were being cut up into subdivisions with streets named things like Windsor Lane and Buckingham Court. This was apparently meant to lend grandeur to the rows of modest new houses. The tiny grocery store was tricked out to look like a 7-Eleven, and the tavern's name was changed to the 19th Hole. There were more people, more houses, and more flower beds.

Spot was a great lover of flower beds. In fact there was nothing he enjoyed more than lifting his leg on a flower bed. The results were apparently deadly. Complaints began to multiply. We were caught in a clash of cultures, like the battle for open range, the cowboys versus the townies. We and our old egg-farming, truck-farming, and hobby-farming neighbors with our free-ranging dogs had to be tamed by the new people, suburbanized. Ordinarily my father would've told these whiners to kiss his Old South behind. But one of the neighbors who complained was Mrs. White, a very attractive and well-made woman. The day she came by for a visit to complain about Spot, my father became astonishingly gracious and charming.

So Spot was confined to the poultry yard. For some reason, we didn't have any poultry at the time. The fence around the poultry yard was made of heavy hog wire and six feet high, high enough to keep out foxes and other chicken- and duck-stealing critters. I saw my friend and companion imprisoned like a felon. All day long he leaped against the fence, yelping and howling. From the time he was a pup, Spot had gone wherever he wanted and done whatever he pleased. I could only imagine his desperation.

Now, if I had the wings of an angel,
over these prison walls I would fly.

When my brother and I stepped off the school bus that first afternoon, Spot was not there to greet us. But we could immediately hear his distant howls from the poultry yard.

We went out to comfort him. He seemed crestfallen, bewildered. He must have felt betrayed. The next afternoon we again tried to console him. On the third day he was expecting us. And he had a plan. When we approached the poultry yard, Spot was nowhere to be seen. When the gate was swung back, he appeared from behind a little grape arbor and made a desperate dash for the opening. We both lunged for him and wrestled him down. It took all our strength to hold him and get the gate closed. I think about Spot and his imprisonment now when I see those bumper stickers that read, "If you love something, set it free." We didn't dare.

My brother and I were in school most of the day. We were subjected to only a couple of hours of Spot's lamentations because he stopped howling at sundown. My father usually arrived home late. He spent a long day at the office capped off with an unhurried visit to the 19th Hole. By the time he got home, Spot had long since yelped himself into exhaustion. I suppose my mother was the one who suffered most. She said Spot yapped and wailed more or less all day long. She began to have that look most of us have in our driver's license photos.

My parents started having odd whispered conversations. If my brother or I entered the room, they'd immediately hush up. It didn't take my brother long to guess what they were whispering about. "I think they're talking about getting rid of Spot."

"No!" I protested. "They wouldn't do that!" I was nine.

Sure enough, that evening Mother tiptoed into the subject by posing a question. "Don't you boys think Spot would be happier someplace where he could run free and not be penned up?" So it was to be all about Spot's happiness? How could a child argue with such logic?

"Like where?" I blubbered.

"Maybe with a family farther out in the country, a place where there aren't many neighbors."

I could've reminded my parents about the great fortune Daddy's very rich and prominent friend had offered for Spot. I chose not to for obvious reasons. My brother was not half as upset about the prospect of losing Spot as I was. He was older than I by three years and at the time had a crush on the girl who played the other string bass in the school orchestra. They read from the same sheet music.

Coonhounds have very loud "voices." This quality has been encouraged through selective breeding for hundreds of years. It enables hunters to know from some distance away when a dog has "struck" the scent or "treed" the quarry. A coonhound's baying can carry for miles. Soon we began to get a different kind of phone call. One of the new neighbors in one of the new houses said her aging mother couldn't rest for the racket. Another lady said her baby was deprived of his nap. One man called and said he was trying to decide if he should shoot the dog or shoot himself. Finally Mrs. White's husband called. "Forget the damned flower beds!" he said. "Forget everything! Just turn that howling moose of yours loose!" It's a good thing my brother answered the phone. My father wouldn't have tolerated being shouted at.

That was all my family needed. When we turned Spot out of the poultry yard, he took off running across the fields until he was out of sight. Then he hid out somewhere for several days. I don't know where. I'm sure he thought that if we'd locked him up once we just might do it again. Finally he returned sheepishly, with a great yearning for a bowl of dry dog food covered in chicken gravy.

After that Spot wandered far and wide. We never heard any more complaints about flower beds. But I don't think it was because he'd changed his habits. There were, however, a few complaints about surprise litters. What was supposed to be the result of pairing a prize female with a pure-blooded champion of her own breed sometimes turned out to be big puppies with long ears and little black spots. But those complaining owners didn't know how lucky they were. It goes without saying the pups had loud voices. But I also have no doubt they all grew up to be strong and brave . . . and excellent dancers.

An L.A. Beagle

BY SUSAN GREGG GILMORE

My beagle is a purebred, white-tipped-tailed, big-eared baby. Sure, he looks like any one of the pack of rabbit hounds my uncle Ed kept penned beside his trailer in East Tennessee all my growing-up years. But he's nothing like those real hunting dogs.

My Bubba is a sensitive boy. He may have a keen sense of smell, perfect for nosing through the kitchen garbage, but he is not scent driven. He is fear driven. He has a delicate, anxious spirit. He's afraid of the rain, crowds of people, even our twelve-pound Chihuahua mix, who claimed her place as the alpha dog in our family pack of two long ago. Then again, my beagle is from L.A.

I had been living in Pasadena for nearly ten years when I decided I needed a beagle. Needed, not wanted. I was missing home and longing for all things Southern. Every sensory receptacle in my brain was flooded with thoughts of overly sweet tea, buttered grits, banana pudding. All clichés, yes, but your memories are colored that way when you're more than two thousand miles from home.

I spotted an ad for a beagle pup and immediately pictured myself lingering outside my uncle's chain-link pen, begging him to take me hunting just so I could ride along with the dogs in the back of his pickup. Begging got me nowhere then. This time, my husband only asked that we name the puppy Jeff Lynne, after the lead singer of ELO. I knew I was going to break that promise even as I was making it–not typical of my behavior, but I was homesick something bad. So Bubba was procured under false pretenses, and in an L.A.-karma way of thinking, I may very well deserve this psychologically challenged dog.

I paid five hundred dollars for him even though I knew the classifieds back home asked no more than fifty. Some litters were offered up for free. Every dog I had ever owned had come from a shelter, and as I wrote the check to the breeder, I promised to donate more often and more generously to the local shelters. (Reminder to self: Write another check.) But my dog's pretty papers came with the California state seal emblazoned on the top in impressive gold-colored ink. My beagle was certified, authentic, even though I lost the papers within weeks of bringing him home. Of all the puppies there the day I picked him up, Bubba was the one that came right to me, big ears nearly dragging across the ground. He chewed on my shoelace and whimpered and fussed, waiting for me to lift him. At the time, his needy nature was cute.

During the forty-five-minute drive back from the Valley to Pasadena, he cried. He cried and cried . . . and then cried some more. I should have realized then that he was a special dog with special needs, not the rugged outdoorsy type you expect of a true hound.

But I excused his behavior. "He's so young. This is traumatic. He misses his mama," etc.

When we got home, I set about reinforcing our fence, certain that if Bubba were to dig his way out, he'd run free, far and free, tracking some animal scent to the point of exhaustion. A few days later, I let him out to play and within minutes he did escape. But he ran straight to the front door, crying for someone to let him back into the house.

Oh, the crying. He cried on walks, wanting me to carry him in my arms. He cried at bedtime, wanting to sleep beside me. He cried if I left the house. He cried when I came home. He still cries if he's left alone outside, and is darn near catatonic when a thunderstorm sweeps by.

In hopes of emboldening him, I watched lots of episodes of Cesar Millan's *Dog Whisperer.* I read books on training dogs and dog behavior. I read to Bubba, anything to calm him. A friend even suggested I call a pet psychic she'd read about, as if coming to terms with Bubba's past lives would bring him (and me) some peace.

Thankfully, he improved some with age. He grew to love his walks (and my husband's over-the-calf dress socks, which the vet kindly returned after Bubba coughed them up during a routine visit). He loves to sit on the porch and watch a light sprinkle, although he's still terrified of thunder. He likes to jump in the car but doesn't like the car to move.

So when we moved back to Tennessee, Bubba chewed off one of his rear toenails before we even crossed the California state line. We stopped at a Walgreens and bandaged his bleeding paw and found a vet outside of Kingman, Arizona, who gave us a big bottle of pills and promised they'd get us to Tennessee. He also suggested we consider a daily dose of Prozac. Bubba's toenail grew back, although it doesn't look the same; it's thinner and weaker than the others. We tried an antidepressant regimen but were never sure it had any effect.

Once we got to Nashville, we had no choice but to rent a small apartment for a couple of months until our house was ready. I had arranged for Bubba to stay with a family friend on a working farm about forty miles outside of town–fifty beautiful acres with ponds, horses, cattle, even one very large picturesque red barn. My husband assured me that Bubba would be in paradise.

It was cold and snowy that February day I drove my beagle out to Lebanon, Tennessee. Mr. Duncan greeted me dressed in overalls and wearing thick black-rimmed glasses fixed low on his nose. He looked kind and wise, and the soft twang in his voice was charming. I pictured Bubba curled up at his feet during the evening, both of them growing sleepy by a crackling fire, exhausted from a day spent walking the land.

I unpacked Bubba's bed, his chew toys, his favorite stuffed goose, his special food, his special harness, his medications, and his Walgreens prescription card. Mr. Duncan smiled as he stacked it all inside a nearby shed. He explained that Bubba would be sharing a pen with Molly, their little female beagle mix. They even had a second doghouse ready in the pen just for Bubba.

"Outside?" I asked as I held my hands in the air, palms upright, catching snowflakes as proof that it was too cold for Bubba to be unprotected from the elements.

"Outside," he said, and smiled. Mr. Duncan assured me that the dogs would be fine. After all, Bubba would surely wander into Molly's house if he got cold. They'd cuddle up together. Besides, he had padded the doghouse floor with wool blankets and had even tacked another blanket across the opening for protection against the wind. Of course, if it got down to ten, he promised me, they'd bring the dogs inside.

"Ten," I repeated. "Freezing is thirty-two degrees. *Ten?*"

Mr. Duncan smiled. "Ten."

Bubba kept close to my side at first, but then with his nose to the ground, and ignorant of the conversation about his housing conditions and the week's arctic forecast, he wandered off. He kept his head low as new and strange scents led him up a small slope, which is why he didn't see the electric fence and ran yelping after the current buzzed the top of his head.

Mr. Duncan laughed. "Don't worry, hon," he said. "We'll make a real dog out of him."

This time, I was the one who cried, all the way back to Nashville. I called every day to check on him. I watched the weather closely, and I cried every time the temperature dipped near freezing. "He is not a country dog," I told my husband. "He's from L.A."

Every day Mr. Duncan reassured me that Bubba was doing fine. He and Molly had been shacking up together, spending the short winter days huddled side by side in their wool-blanketed nest. By the time I picked him up, in the middle of March, the temperatures had turned warm and the grasses bright green, and Bubba was happy. He might even have been in love.

He changed on that farm. Maybe the fear was shocked right out of him at the start, or maybe he got in touch with his roots, or maybe Molly opened his heart and fortified his courage. I'm not sure, but I know he's different–calmer, more relaxed, more at ease in his own coat.

When I stepped out of my car, Bubba ran over and jumped and licked and did everything he could to reassure me he was not mad at me for abandoning him in the cold like that. Somewhere out on that Tennessee landscape, he became more of a real dog, just as Mr. Duncan had promised. I'm not sure I'd take him rabbit hunting. But at least he's no longer afraid of the rabbit.

Believing in Chance

BY HUNTER KENNEDY

About a year ago, I bought a dog off the Internet. This ten-week-old English springer spaniel had been bred by a kennel in Tivoli, Texas, that specializes in field-bred spaniels. Having done some research on the breed, I knew I didn't want a show dog, but rather a real hunting dog that was smart and well tempered and had a nose for birds. The mother was the number three gundog in North America, and the father had his own share of blue ribbons, so I knew there was a long line of buyers eyeing the litter. After a couple of hasty conversations with breeder David Jones, I bought a male pup sight unseen in a transaction that could only have taken place in the early twenty-first century. Jones sent me a few fuzzy images of the puppy he'd picked out for me via e-mail, and then I called the breeder on my cell phone from the airport to confirm the little guy's safe arrival via Delta. Now that my randomly selected English springer spaniel was barking in the backseat, Jones wanted to know what I was going to call him. "Chance," I replied. My grandfather had taught me to hunt, and I had always wanted to name a dog after his springer Chancellor. In this particular case, I thought it appropriate.

To be fair to my grandfather, he did not teach me how to train a dog. Nor did my dad, who is more of a cat fan. In fact, I had not owned a dog in twenty years, and had no idea exactly what was required to turn this bird dog into a "good dog"–a quality required of all canines with which one would trouble to associate. I wanted other people to associate with me and Chance–mainly my wife, who was not keen on the idea of getting a dog, but who had allowed herself to be persuaded by the promise that English springer spaniels were "pretty mellow." I neglected to add that this mellowing occurred around age ten; until then, I hoped we were up for the challenge.

Advice came from all quarters on how to train my new hunting dog. The breeder sent me a few more e-mails. Friends and family showered us with dog books. Even common strangers peppered me with advice. Dogs are like barbecue–everybody is an expert on the subject of what makes for a good one. My favorite piece of wisdom came from an old quail hunter who said, "You just need to talk to your dog a lot." Chance and I talked all the time. I found myself repeating basic phrases over and over until I wondered if the college kids next door thought I was a raving idiot. Chance would look up with his hazel eyes, cock his ears quizzically, and go back to doing whatever it was he thought natural. This included barking all night, scraping the paint off the kitchen door with his front paws, and putting a doughnut hole in his new dog bed. But anytime he did something right, however coincidental,

I would repeat the words "Good dog" and give him a treat. To a certain degree, this positive reinforcement worked. Chance grew to understand that he would be rewarded if he did certain things. Unfortunately, it took him a little longer to realize that other actions were to be avoided.

This included raiding the mailbox. My mailbox was installed by the previous owners of the house, and they obviously didn't have dogs because it had no lid. Anytime a large bundle arrived that didn't quite fit inside the box, Chance assumed the mailman had dropped off a new toy. The first few times he got into the junk mail, I gave him a stern talking-to and cleaned it up. Then, one day in April I came home to find my tax returns scattered across the courtyard in little bits of confetti. Chance was gyrating in the middle of the pile in a one-dog parade. Needless to say, our communication swiftly moved beyond verbal. Chance was momentarily chastened, but did not stop there. He shredded a fat packet of closing papers mailed out by my lawyer. After we finally got a lid for the box, he ate a six-foot-tall Japanese magnolia limb by limb.

Every evening when I came home from work it was a new surprise. He pruned the lower half of the fig tree and frequently dined on my wife's favorite gardenias. A half-dozen rose bushes became springer salad. There were frequent raids on the recycling container, which was impossible to dog-proof. Chance spent his idle afternoons tearing aluminum cans into bright ribbons. Plastic bottles were broken down to a molecular level. And once I came back late from work to find him chewing on the jagged shards of a margarita bottle. I pulled him away from the pile of glass and quickly checked and rechecked his mouth for blood. Not a cut. That was the moment I finally understood the true meaning of the term *dog years*: It does not apply to a dog aging seven years for every year of a man's life; it means that a dog makes every year feel like seven—especially if he is not a particularly good dog.

It took me a few months to realize that Chance had been set up to fail. This bird dog had been bred for energy and stamina, and he could run all day—if asked. The problem was that he frequently wasn't. Instead of being given a broad field to patrol, he found himself in the cramped courtyard of a Charleston single house, which, like many houses in the city, has no backyard. Thus the access to the mailbox and the recycling bin. Thus the garden waiting to be dined upon each afternoon. He could no more articulate what he needed than I could. He barked. I commanded. And in those bittersweet moments when there was a fleeting connection of language between beast and man, when Chance divined what it was that was required, he was told the same two words until it finally began to click.

He really is a good dog. It doesn't matter that he set an Orvis Company record by gnawing through their guaranteed chew-proof Dog's Nest in fifteen minutes; or that the only reason he graduated from obedience school was that they had already written his name on the diploma. Despite all the bad things Chance has done in his brief career, we have spent so much time together in those laughably one-sided conversations that somehow he really has become quite a good dog. Maybe he just likes the sound of those two words. Now, when I'm washing the mud off Chance and he patiently waits for me to step away before shaking out the water, I tell him he very much is.

Emmett & Me

BY JOHN ED BRADLEY

I never wanted a dog. I preferred cats. But it was Christmas and he was a gift and I didn't know how to tell her I couldn't take him. She said the breeder was a lady in town. There'd been an ad in the paper: English bulldogs, sired by a champion, $450 each. She'd picked out the best one. "How much did you say he cost?" I had a hard time keeping my voice from shaking. She didn't ask me to reimburse her, but I wrote out a check anyway. We were both so poor it didn't make any sense.

I named him Emmett. That first night together, I put him in a cardboard box with a blanket on the bottom. He cried so much I had to take him to bed with me. "You miss your mama?" I said. "Is that what has you so upset?"

I held him against my chest and let him hear my heart beat, and he calmed down after a while. He slobbered, though, and drool got on my sheets and glued my fingers together. He tried to suck one of my nipples and I had to push him away. When he finally went to sleep, he snored like an old man with a deviated septum. Even worse was his intestinal distress. It got so bad I had to get up and go outside for some air.

I noticed how he'd whine and paw at his ear, and how he'd give you a look if you tried to pet the top of his head, so I took him to the vet. The man tried to be gentle about it. Emmett had a bad ear infection, he said. His tubes were blocked and forming scars. "Your dog's deaf," he said. "But it's not just that. If we don't stop the infection, it's eventually going to kill him."

"Give him some medicine," I said.

"I will, but he needs surgery, and not the kind I'd attempt here."

I took him to the LSU vet school in Baton Rouge. When it was over, they carried him out to me wearing a big halo fitted around his neck. It kept him from scratching his ears. He was so excited to see me he peed on my pants and best Sunday shoes. I didn't have enough money to pay the bill. I had to use one of those cash-advance checks a credit card company had sent me in the mail.

It's hard enough to communicate with a dog that can hear. Imagine what it was like trying to train Emmett. You could clap your hands together in front of his face and he wouldn't flinch, but I worked with him, I worked hard, and over time he started to obey sign language. If I caught him drinking water from the toilet, for instance, all I had to do was stomp on the floor and point at the door to get him to stop and leave the room.

You could bathe him in the morning with the garden hose and some baby shampoo and by noon he smelled bad again. You could freshen up his water bowl three times a day and he was still going to leave ropes of saliva on the furniture. He'd break wind at the wrong time, too, like when you were just sitting down to supper. And if you had guests over, you could count on him humping their legs. He'd hop right on up there and go to town, a look on his face that let you know he was in the murderous grip of something that even he did not understand.

I knew he needed a girlfriend, so I called around to see if anybody had a female bulldog they wanted to breed. After a few weeks a man showed up with a girl dog chained to the back of his pickup. "You got your papers?" I said to the man. He didn't say anything; he just looked at me. "Emmett's registered with the American Kennel Club. I'm sorry for the misunderstanding."

"You think he's too good for Madonna?"

"No sir," I answered. "Bring her on back."

When they were finished, Emmett slept for two days. I wondered if he'd slipped into a coma. He'd given his best effort and made me proud, but I was even more impressed when, nine weeks later, Madonna delivered a litter of eleven puppies, more than double the average for the breed. To celebrate, I grilled a couple of rib-eye steaks and fed one of them to Emmett. They were discounted manager's specials, a few days past their expiration date and turning color, but Emmett didn't complain. I popped a beer open and poured it in his bowl; then I had one myself.

On weekends we often went for drives in my truck. Most days we traveled the country roads, but other times we visited the French Quarter in New Orleans. If it wasn't too hot outside, I put him in the bed and let him feel the wind. He liked to perch himself on the wheel well and have a look at all the people on the sidewalks. They'd run over and pat the top of his head, his little bun-shaped tail shaking. Tourists fed him Lucky Dogs, beignets, and jambalaya. They let him sip from their go cups. The strippers and transvestites on Bourbon Street, the art dealers and antiquarians on Royal and Chartres–it seemed everybody had to run out and say hello.

I was often mad at him for one disappointment or another, like the time he ate a frog, came in the house, and threw it up on the carpet. But I can also tell you there were days when it seemed he was the only friend I had in the world. I'd sit on the back stoop and read my bad reviews to him. A hurricane hit the town and I sat holding him close in the bathtub, waiting for the walls to blow away. When my girlfriend and I were fighting, I'd take his face in my hands and whisper in his dead ears about how I had him and didn't need a woman anyway. "You're lucky you found me," I told him, even though I knew that I was really the one who had it good.

He died on Thanksgiving Day 1995, after a family dinner at my mother's house. He was only six years old. He'd been sleeping longer than usual and I went to wake him up. I gave him a shake and said, "Let's go, you randy, foul-smelling beast," but he was cold, his body rigid. His chin rested between his front paws, and his eyes were closed. We figured he must've had a heart attack in his sleep after eating turkey and cornbread dressing with us, and after playing for hours with my nieces and nephews, who'd pretended he was a horse and taken turns riding on his back. "Emmett," I ran into the house screaming. "He's dead. Emmett's dead."

When things settled down, my brother and I went to the shed and came out with shovels. There was a grassy place on the side of the house where we'd buried our childhood pets. This was where we put him.

"Why don't you get another one?" people asked me.

"You know why," I answered.

About ten years ago a friend called and told me she had something in her office that I needed to see. This friend is a professional art conservator; she repairs damaged paintings and pottery for a living. A client had bought a large ceramic bulldog at an auction, but the rare life-size antique, made in France in the 1890s, had been damaged in shipping. The client had filed a claim with her insurance company, and now the company's adjuster wanted to sell the dog.

"It looks just like Emmett," I told her when I arrived at her studio.

"You want to take him home with you?"

I keep him in a spot next to my favorite chair in the living room. I pat him on the head every time the doorbell rings and I go to answer it. As much as I enjoy owning the dog, I've never for a minute been fooled into believing he was Emmett. He doesn't place his paws on the top of my feet when he wants my attention, and he's never once licked me so much that I've needed a shower.

I still prefer cats, yet I dream about Emmett. He's outside barking at a squirrel even though I moved from that house fifteen years ago. I hear him snoring in the other room and when I go there the room is empty.

It's hard, all right. But I suppose that's what you get for loving somebody.

CHAPTER 2

Into the Field

The Trophy Huntress

BY JONATHAN MILES

Used to be, when you picked up a dog from the Lafayette County, Mississippi, animal shelter, you had to state your intentions. On the adoption form was a checklist of reasons for wanting a dog, requiring you to mark one: Was it *companionship* you were after, or maybe *protection*, or was it *hunting* or *breeding* or *gift* or the melancholy (if also vaguely callous) *replacement for a dead pet*? For a long contemplative while I stared at that checklist, then at the mange-splotched pup I'd just chosen, and then back again–long enough for the shelter attendant to inquire, gently, if I maybe needed help with the form. I was twenty-three that summer, and–having just moved into a 12-foot × 30-foot cabin amid 350 acres of hill-country hardwoods, where I dimly intended to confront the bone truths of existence by living off the land–I was saddled with far more ambition than sense. *Hunting*, I marked. The attendant glanced at the form, then at the yellowish pup. Best anyone could ever figure, she was a mixture of retriever and collie with possibly a smidgen of coyote thrown in; years later, when reporting jobs took me to South America, I was finally able to pinpoint her breed as Latin American Street Dog (*Canis wantus empanadas*). The attendant gave me a weak smile. "Good luck with the hunting," she said, not without irony.

I called her Julip, after a character in a Jim Harrison novella whose name derived from a "mixture of a flower and a drink." A swell name for a bird dog, I thought. Julip's bird-dog career, however, lasted only a few months. Aside from her mongrelized disposition and an inept trainer, her main problem was guns. It wasn't that she was gun-shy. Instead, she loved guns *too* much, but for the same reason little boys love guns: because they go *bang*. While she was able to find and flush quail, the moment a trigger was pulled, it was all over. A berserk tail-whipping glee would overcome her, all bird thoughts vanishing as she'd come charging back at me, squealing and yapping and all but knocking me backward, as if to say: *Make that noise again! Now!* For Julip, birds were not the object of a day afield. Bangs were.

My landlord was an elderly Virginian who visited once a year to collect back rent and take stock of the property. The latter task entailed walking the woods for several hours, armed with a 12-gauge he'd fetch from the trunk of a dilapidated Buick that was otherwise stuffed with fifths of Aristocrat vodka. "In case we see something," he'd say, with a disturbing lack of specificity. Near the end of her sporting career, Julip and I tagged along on one of these annual hikes. My heart thumped with pride when she flushed a covey of bobwhites at the edge of a vast blackberry thicket. My landlord, with a display of reflex and agility I found startling considering his age and vodka consumption, swung his shotgun skyward and fired off two impressive if unsuccessful rounds. Julip, of course, immediately assailed him, with a cringe-inducing performance of somersaults and yips that no amount of stern, embarrassed correction could halt. She even tried licking the shotgun barrel. Squinting, my landlord appraised her with a curled lip, then turned to me gravely. "I don't think that dog'll hunt," he said. Flattened by truth, I just nodded.

She *did* hunt, as it turned out–just not with me. Oftentimes I'd return home to find a half-eaten rabbit carcass on the floor, or sometimes on the bed. One time I awakened to the sight of a bucktoothed squirrel staring me square in the eye, right beside me on the pillow. Also beside me, farther down in the bed, Julip was delightedly chomping on its lower portions. (So many pillowcases I lost to bloodstains. A poor unshaven bachelor living Unabomber-style in the woods cannot have bloodstained pillowcases if he desires female company; this is what's called a deal killer.) Frogs were another common prey. Also snakes. I once caught her with the foreleg of a deer, but I'm pretty sure that was a salvage operation. It was, I see now, a decidedly feral existence–on both our parts. Back then I was drinking too much, drifting through minimum-wage jobs, bumbling in and out of messy love affairs, and intermittently having the time of my life. Beside me through it all was Julip, my sidekick, my confidante, my partner in excess. Once, at a pig-cooking, she cadged so much pulled pork as to immobilize herself. I discovered her lying on her side in the grass, her belly swollen to the size of a soccer ball, panting but grinning. An anxious call to the vet revealed that an overdose of pulled pork is almost never fatal, a fact I have come to find valuable in my own life. I had to seek medical care for her after another party, this one an all-day event involving a cauldron of chicken stew cooked over a campfire; she'd spent all day and night biting at the sparks thrown off by the fire (sparks were also among her prey), which resulted in a hacking hangover. The vet said she was the only animal he could remember treating for smoker's cough. She once wandered off to a backwoods party where some acid-dropping Ole Miss students spray-painted her with glow-in-the-dark fluorescent paint; for days she was a greenish phantom, a four-legged glow stick roaming the woods.

Eventually, when she was seven and I was exiting my twenties, we had to get civilized. I understand this to be a common fate. First came marriage, and with it three more dogs and a house that contained more than one room, then a real job in New York City, followed by children Julip never once tried to eat. For kicks I would take her to Central Park to wow her with the sight of the giant, tame squirrels there; I am certain that afterward, when she dreamed, it was of an unleashed, belly-swelling massacre on the edges of the Great Lawn. Outside of dreams, however, her retirement was placid: a long happy bangless nap time. Her name was one of my eldest son's first words. She liked licking the top of his head, to make him giggle and kick his fat jellied legs.

Dog stories, of course, are like any biography: There's never a happy ending. In her twelfth year Julip developed lymphoma, and steadily dwindled. By this time we'd moved out to the country, in upstate New York, where she'd happily if less vigorously reignited her campaign to rid the world of small mammals. I'm inclined to believe that the worst she suffered, toward the end of her life, was the growing indifference of our local rabbits, which grew bolder and bolder as Julip waned until, near the end, they'd munch on clover within fifteen yards of her. More than once I burst outside howling, to scatter them for her dignity. The look she'd give me afterward was less appreciative than critical, as if to say: You'll never nab one *that* way.

One Saturday morning, in the maple-red autumn, while my wife was traveling on business, I packed my three-year-old son and one-year-old daughter in the truck to haul Julip to the vet for her weekly chemo treatment. The vet, an ancient man with a long gray wizard's beard who in a past life had been a zoo vet, always called her "beast," and as the needle found its vein he stroked her ears and narrow muzzle. But the shot was too much for that beastly but wearied heart of hers to endure. She died in my arms as I was carrying her back to the truck. The vet and his wife entertained my children with stories about all the grand animals in the photos on the walls–elephants, lions, etc.–while I sat on a plastic chair in a back room and wept and wept and wept.

The next afternoon, while driving home from the grocery store, I saw a coyote bound across a lonesome back road and then across a cattle pasture, running as smoothly as a panther, in what was clearly hot and beautiful pursuit. I didn't think it was Julip, or a mystical sign, or even a significant coincidence, but I also didn't know I was stopped until I glanced in my rearview mirror and saw four cars idling behind me. It was just a bitter comfort, in that moment, to see something hunting out there.

Poodle Power

BY LOGAN WARD

The World's Tiniest Bird Dog arrived December 1974, in the wriggling toe of a Christmas stocking. Dad laid the glittering red-felt pouch beside the hearth, and out tumbled a fur ball the color of a Hershey's Kiss and only slightly bigger. Along with our heaps of toys, here was a toy poodle. I'd never seen a dog small enough to perch on a man's palm, literally. As my father cupped the sniveling brown powder puff, none of us could have imagined the pup was meant for anything but lap naps.

We named him Tom–Thomas Thumb Ward. To my eight-year-old ears, Tom seemed as good a name as any for a dog, though I came to realize both moniker and dog were eccentric. Our up-state South Carolina neighbors had more conventional canine companions: Princess, a boxer who greeted visitors with over-joyed snuffles, a clatter of claws on hardwood, and kidney-bean twists; and a gray-bearded black Lab named Blue, an old gundog who roamed the wooded lots until his master, a private man who worked for the forestry department or maybe Fish and Game, bellowed him home: "Heyyyyyy, Bloo! C'monnnn, Bloo!"

Like most small-town doctors' wives back then, my mother played tennis. There were always tennis balls bumping around our station wagon's way-back or gathering dust bunnies in utility-closet corners. Though none of us remember the first fetch, at some point during his early development, when his mouth was big enough, Tom plopped a felt-covered orb at our feet and whimpered. Thus commenced Ball Fever.

Tom's exploits began in the TV room. He was an indoor dog, after all, small, not a shedder. As my brother, Bill, and I whiled away hours in front of the boob tube, we sent Tom scampering, bouncing tennis balls off walls, arcing them behind the sofa, flinging them under chairs, and ricocheting them down the hall and into the dining room.

Tom's tennis-ball obsession became a parlor trick. We invited friends. Come watch the tiny poodle fetch balls bigger than his head! Soon the panting expectancy, yipping, and dropping and redropping of balls into laps grew tiresome. We tucked Tom's ball inside cabinets, atop the mantel, or behind books lining a tall built-in shelf. Tom was no quitter. He'd hop like a mountain goat from floor to chair to shelf, or paw at a couch cushion as if he were digging a hole until we feared he might.

No house could hold Tom, lapdog or not. Our games of fetch moved outdoors. I'd pump my arm, faking a throw, and while Tom sprang off in that direction, I'd spin around and hurl the ball into the woods or distant azaleas. But it was just a matter of time before Tom trotted back, head held high, not out of pride but to balance the heavy ball.

Our dad thought it only natural to take his little "Tom Tom" on a dove shoot. Centuries ago, large poodles were bred to hunt, but their reputation took a U-turn after they became coiffed and pampered pets of the French royal court. More than a century ago, W. R. Furness wrote in *The American Book of the Dog* that if you accused a dog man of owning a poodle, "he repudiated the charge immediately, and felt deeply insulted, as these dogs were deemed fit only for the circus or for mountebanks."

After one hunt, as we cleaned birds over newspapers in the garage, Dad pulled a dove from the pile, gave Tom a whiff, and tossed it into the woods. A brown blur shot after it, the tiny form swallowed up by the underbrush. We heard rustling and then nothing. The tennis ball would be back in thirty seconds flat. We waited.

"Fetch, Tom!" Dad hollered. Nothing. "Tom Tom!" he said through clenched teeth. "Get your little butt back here!"

Tom pranced back with a face covered in feathers, licking like mad to remove them.

"He was all juiced up," Bill remembers. "That dove did something to Tom. He went cuckoo over it."

After that, Dad gutted a dove and stuffed the breast with cotton, lacing it up again with needle and thread and storing it in the crisper drawer of an old beer fridge. After a long day at the hospital, he'd stand in the driveway, firing blank rounds from a starter pistol while tossing the fleshy trainer. Time after time, Tom wobbled back, head sagging, mouth stuffed with bird, struggling to see over the feathers.

That diminutive dog was a quick study, and one day Dad, in his eagerness, decided Tom was ready for a field trial. And why not? Though half as big as a house cat and hardly heavier than a mallard (which made duck hunting out of the question), Tom had the right stuff, even if he occasionally had to stop, drop his quarry, and rest. Size wasn't really the issue, though, at least not for me. Privately, I wondered if we–more pointedly, if I–could handle the embarrassment of entering a dove field trailed by a pint-size poodle.

In the field, we discovered another problem. We had driven a dozen miles to an old dirt farm Dad had bought for weekend dabbling. The land was long fallow, so birds were scarce, but we packed the stuffed trainer in a cooler full of cold drinks. Tom ignored the shotgun blasts and went straight to work, bounding through weedy thatch and reemerging minutes later with a big gray bird in place of his head. The problem was cockleburs. With every pass, those curly locks snatched up the spiky seedpods, which worked themselves deep into his long, fine hairs. Pulling cockleburs off your tube socks is one thing. Back home, Tom squirmed and our fingers bled as we picked the burs one by one out of his matted coat.

What Tom needed, Dad decided, was a hunting suit. This was not the most practical solution. Dog apparel was only beginning to catch on. True, the trendsetting Tom owned a navy-blue crew-neck sweater, but I doubt there was a store in the world that sold a poodle hunting suit, at least not in Tom's size.

Undaunted, my dad got his hands on several yards of beige Ultrasuede and tracked down a seamstress willing to custom-tailor the tiny ensemble. Ultrasuede–the world's first synthetic microfiber–was all the rage, at least among the ladies of our mill town. They bought deeply discounted yards of it straight from the plant. My mother proudly wore the forest-green Ultrasuede vest she had made to Ladies Auxiliary luncheons and other dress-up affairs. Tom was less enthusiastic about his getup. The design was fairly ingenious, a bit like today's canine saddle coat but with leggings that tapered into fitted booties. A brimless cap made from the same beige Ultrasuede tucked around his head and fastened below the throat. The struggle to dress him must have seemed to Tom like some strange new torture. The only part he could not abide were the built-in booties. Dad knew to pick his battles and snipped them off just below the dewclaw.

On the morning of Tom's first public hunt, Dad steered off the blacktop, and we jounced toward the edge of a stubbly cornfield, where a clump of camo-clad men and boys and their dogs were about to meet Tom, a perky poodle dressed like Oliver Twist in knickers and newsboy cap.

"Hey, look at that little dog!" a man said as we approached, lugging dove stools heavy with #8 shot. "What are you gonna do with that thing?"

"We're going to hunt," Dad said.

"With that sissy dog?" he asked. Other men laughed.

"Don't you worry about this little dog," Dad said. "You'll see."

The chatter continued, but soon we split up and spread out around the field's perimeter in search of cover. Tom, hyper by nature, darted off. "Tom Tom!" Dad called. He returned but kept weaving around, sniffing things and yapping at the black Labs and spaniels. Tom could retrieve, but could he heel?

He could not. Still, between Dad's clenched-teeth rebukes and Bill and I snatching him up whenever he scooted past our stand of cornstalks, the three of us managed to keep Tom under control. The gun blasts helped focus his attention.

The rate of incoming birds picked up. Dad fired at an approaching trio, and one fell, landing nearby. He pointed and spoke—"Back!"—and Tom zigged and zagged after it, his Ultrasuede suit rustling the corn stubble. When he returned, bird in mouth, I relaxed a bit, though the man who had branded Tom a sissy hadn't noticed.

Tom got another chance to prove himself. Though I don't remember the details, according to my father, who is now eighty, the same man downed three doves in quick succession. Tom slipped our grasp and bolted toward the place where they fell. As he wobbled back to us under the weight of a bloody dove, the man poked his head above his cover and yelled, "That dog's taking my bird!"

"He's a bird dog," Dad said. "Don't worry. You'll get it back."

Tom fetched the man's other two doves, and later, as we stood beside the vehicles sucking down cans of Mountain Dew, Tom lapping water from my brother's outstretched palm, the man approached. "Doggone if you weren't right," he said as Dad handed him his birds.

With the hunt over, the beer-sipping men made a fuss over Tom's performance, though I doubt any of them were sold on the idea. To them, he was a novelty.

In all honesty, they were right. If Tom's story were a Disney movie script, he might beat one pedigreed pointer after another to win the National Field Trial Championship. But Dad wasn't looking for a champion. Had he been that committed to the sport, we never would have owned a toy poodle in the first place.

Tom continued to join us on hunts–never learning to heel but always sparking conversation–but he seemed just as content scrabbling across the den floor to fetch his smelly tennis ball. His final years were even more un-Disney-like. His hair thinned, and his breath grew rancid from teeth so infected the vet had to pull them all, causing his tongue to loll out of the left side of his mouth.

I lost touch with Tom when I went off to college. As his time grew short, he remained as sociable as always, even without his teeth. "The day before Tom died we had people over, and he was right there in the middle of it all," Dad recalls.

Today, we live in an age of numbers, of so-called Big Data. I hold no quantitative proof that Tom was the World's Tiniest Bird Dog. Does it really matter? He was small, and that dog did hunt.

Education of a Bird Dog

BY GEOFFREY NORMAN

The turnoff was five miles from Union Springs, a name that doesn't mean much unless you care about bird dogs. In that case it means a lot. Resonates, I suppose, the same way the name Bordeaux does for people who care hopelessly about wine. Union Springs, which is about forty-five miles east of Montgomery in the Alabama Black Belt, is known as the field trial capital of the world. A bronze statue in the town square depicts not the usual Confederate infantryman but an English pointer, standing staunchly with head high and tail straight.

My wife and I came here to drop off my dog, an almost-two-year-old English setter, to be trained, and I worried that neither my dog nor I belonged here. As if we were . . . unworthy. Sending your dog for training in Union Springs is like sending your musical prodigy to Juilliard. But training him myself wouldn't work. It hadn't with any of my previous bird dogs and not for lack of trying.

Bird dogs can be just as addictive as heroin and are slightly less expensive. There was a bookshelf in my office taken up with volumes on how to train your pointing dog, a lot of them with pages turned down and passages highlighted. The only real lesson I had learned, after much study, was that I am no dog trainer. Whatever the talent, I lack it utterly.

I had never met the man who would be training my dog. I had asked someone who knew the territory to recommend a trainer. Told him what I had in mind, and he said he would ask around and get back to me.

He called the next day and said, "I believe I've found the man you're looking for. This fellow used to do a lot in the field trials. Had a good record, too. But he's not doing that any longer. Nowadays he has a small kennel. Only four or five dogs. He works with them himself. No assistants. And he gets very good results. Two or three people I talked to have put their dogs with him and they swear by him."

"That sounds perfect," I said.

"Good. Let me give you his name and number. It's Ramin Jackson." He spelled out R-A-M-I-N.

I called Jackson, told him how I'd found him, hoping my friend's name might buy me some credit, and said I'd like to talk about some dog training if he had the time.

"That would be fine," he said. "It's almost time for lunch. Tell me about this dog. What did you say his name is?"

"Woodrow," I said.

"That's good," Jackson said, without explaining why.

"He's a setter. And sort of a soft dog."

"Yes. Most of them are."

"But very game. From a real good breeder." I was selling hard.

"Around here?"

"Montana," I said, hoping that wouldn't disqualify Woodrow in Jackson's eyes.

"There's some good ones out there."

We talked for twenty or thirty minutes, and he asked what, exactly, I hoped for with Woodrow. Steady to wing and shot? Retrieve downed birds? Back another dog's point? Was I ready for the possibility that no amount of training would make him into a bird dog?

"Yes," I said.

"Well," Jackson said, "then I'm looking forward to meeting Woodrow."

Not me. Woodrow. I liked that.

Now the day had come, and I was at Jackson's place to meet him and leave Woodrow to his care and training. Jackson was cleaning out kennels when we pulled in and parked next to a small barn. He walked over to greet us. He had a brilliant, genuine smile, and his clothes seemed far too clean and well pressed for the work he was doing. With introductions out of the way, Jackson's focus turned to Woodrow, who was by now out on the grass, moving carefully around the property as if he might want to buy it. "Fine-looking dog," he said. "Holds his head real nice and high. I like the way he moves, too. Yessir. He is a fine-looking dog." Said in a way to convey the old, old truth that looks ain't everything. And sometimes they ain't nothing.

"Course we won't know about him until we get him on birds. You leave him here and give him some time to get used to me. Then call me and I'll let you know how he is around birds."

....

The first call was discouraging. "Woodrow doesn't seem to care much about birds," Jackson said. "But it's still early. Let me try a couple of things. Call me again in a week or so."

Well, I thought, if he won't make a bird dog, he'll still be a fine pet.

After another week, I called again.

"I am so proud of my man Woodrow," Jackson said, his voice sounding especially musical. "And you're going to be proud of him, too. Why don't you come up and visit in another week or two? I'll have him ready to put on a show for you."

When I arrived, Woodrow was waiting on the grass outside his kennel, delighted to see me. So much so that I wondered if, in all the excitement, he might not remember what he'd been taught.

But Jackson got his mind back on business with one simple command. "Here, Woodrow," he said, and that was enough. Woodrow stood by Jackson's right heel looking more purposeful than I'd ever seen him. Like the boy who'd gone to boot camp and was now a man.

"I've got some birds put on that little hillside over there," Jackson said. "Let's go watch him work."

Woodrow stayed at heel until Jackson released him, saying, "Hunt birds." Woodrow tested the wind and went to work. It was, for me, a joy to behold. For Jackson, probably, just another day at the office.

Woodrow coursed the broom sage on the hillside and caught scent. He tensed.

"Careful," Jackson said. Woodrow took another couple of steps and locked up.

"Look how nice he holds his head," Jackson said. "Pretty, isn't it?"

"Beautiful." We walked in. Woodrow did not move. Not when we moved past him and not when the bobwhite got up in a blur and flew off at an angle and not when Jackson dropped the bird with the little 20-gauge he was carrying. Not until Jackson touched him and said, "Hunt birds."

Woodrow went out to the dead bird, mouthed it, then dropped it. "He'll pick 'em up, but he won't retrieve them. Not yet."

Woodrow worked another four or five birds. One of them hit the ground running after Jackson shot. The dog broke and ran for the wounded bird.

"Woodrow," Jackson said, calm but firm. He came back, and Jackson positioned him as he had been before the shot. "Whoa," Jackson said. That was it. No shouting. No hitting. No electricity from a shock collar. Just a firm tone of voice.

Woodrow did not move again until Jackson said, "Hunt birds," and slapped him affectionately on the flank. Then he was back to working the wind for the scent of game.

Jackson had worked miracles. "How did you learn how to do this?" I asked. The way you might ask an artist how he'd "learned to paint."

"Oh, I've always done it," he said. "Ever since I was a kid growing up around here. My father did a little farming and we always had a couple of dogs we'd take out walk shooting."

And was there anything that his father had told him that was the key? The secret? The thing that made everything else possible?

"He always said, 'Don't be too hard on 'em. Show a dog some

love and make him trust you and then he'll work for you.' I've seen him get down with a dog and give him a kiss. Some people think you have to always use force. They'll get on the dog with a shock collar until they've turned him into a blinker. The dog can smell birds and when he does, he just goes on around them. I've seen dogs run away from birds to keep from getting shocked.

"When I get a dog like that, I start out not doing anything. Just let the dog be a puppy again, play with the dog until he gets to trusting me, and work him up slow. You have to get to know a dog, first, before you can train him."

"How do you get to know him?"

"Oh, there's ways. I like to watch a dog for a while. I want to see if he's got a nose for birds. I want to see how he moves. I want to know he can hear me when I'm talking to him. I'll let him get used to me before I put my hands on him. Then I can move my hands along his backbone and see if he's tense. Some dogs are soft and nervous, you know, like Woodrow. I'll go slow with them so they get used to me."

One method he'd used with Woodrow, Jackson said, was to stand him on an empty fifty-five-gallon drum with birds in front of him, or tethered and flying around him, to teach him to be steady on point. If he moved, the barrel would roll and he would fall off. Jackson had used a check cord, mostly, and a shock collar, occasionally, to keep him in range and to remind him to stop cold on the command "Whoa." He had put him in the field with another dog, and when she pointed, he had put Woodrow, who was on a leash at heel, on point to show him how to back. And so on. No magic tricks. Just standard training techniques. Easy to describe and very hard to execute.

"You ran dogs in field trials?" I said.

Yes, he said; that was after the one time in his life he'd been away from Bullock County. When he graduated from high school in the late seventies, he left for Michigan and a job in an automobile factory. But he came home after a couple of years. "I missed this place. And I really missed working with the dogs."

He worked on one of the big hunting plantations around Union Springs in the nineties, breaking horses and handling dogs. Guests who came to shoot at the plantation saw what kind of work he did and asked him to handle their dogs in some of the trials. So for several years in the nineties, Jackson trained champions. He took a runner-up in the Alabama Open with a dog named Native Shadow Mike and was twenty-third in a nationwide ranking of handlers. It was a challenge, he says, but he doesn't miss it. Now he manages some property for an out-of-state owner and works a few dogs in his own kennel. "Only four or five," he says. "That way I get to know those dogs. They're all different, and it's better when you can get to know them and take your time with them. Like with me and old Woodrow, here."

As we talked some more, all my questions essentially came down to this: What's the secret? "The most important thing with dogs," Jackson said, "is patience. A lot of people, they're in a hurry with the dog and they get mad when he makes a mistake and they'll come down hard on him. They can't help it. They don't have that patience. But it makes the dog tense and then he makes another mistake and . . . you see how that goes.

"You need to keep working to get the dog to trust you, like I did with Woodrow. He was nervous when you left him. Didn't know what to make of me. Wasn't interested in birds. So I just let him run and, after a while, he'd come close and let me put my hands on him. Then he'd start to relax a little and I could work with him. I put him on the barrel with no birds around him, at first, just so he'd get used to being still and steady. Takes time. Woodrow is going to be a real good bird dog, now, for you. Real good."

He was, plainly, proud of his work.

"He's a real fine dog. I'm going to miss old Woodrow."

"In that case, I'll bring him back for a refresher."

"You know," Ramin Jackson said, "I was hoping you'd say that."

On Patrol

BY BEN McC. MOÏSE

"Find the bird, Belle . . . you look for it!" I said as we approached a hunter who was standing just inside a row of cornstalks. Minutes earlier, Belle and I had been sitting in a ditch in the middle of a dense stand of wax myrtles. We were just off the edge of a dove field where I was observing some two dozen hunters scattered throughout the field.

My attention was directed toward this particular hunter because I had seen him down fifteen birds, three more than the legal limit, and he had just shot another. Belle, then my sidekick for almost six years, had been in the bottom of that ditch with me whining and squirming with excitement at the sound of each shot. She was raring to go forth and find something.

Her official name was Southern Belle, and she was a Boykin spaniel. Of course, I thought Belle was the prettiest and smartest Boykin spaniel alive. The first of the breed materialized around

1900 in Spartanburg, South Carolina. Legend has it that the first small brown dog was found wandering near a Methodist or a Presbyterian church–though some have said that given how much Boykins love the water, the dog had to be Baptist. Regardless of the possibilities of its quest for spiritual nourishment, that little brown pup was passed to L. Whitaker "Whit" Boykin, who lived near Camden, South Carolina. Boykin spent years developing the eponymous breed by introducing similarly colored spaniels as well as a Chesapeake Bay retriever into the line. In acknowledgment of its South Carolina origins, the Boykin spaniel was proclaimed the state dog by a legislative act in 1985.

In my job as a South Carolina game warden, Belle was my constant companion on land and water, in all states of weather, and at all hours. If you saw me, you saw her. After that much time together, we were on the same wavelength–she could anticipate what I wanted and I knew when she wanted to play, needed a pat, required a pit stop, or simply wanted to be left alone.

Belle had played the "find the bird" game before and quickly got busy sniffing around the corn rows. After introducing myself to the hunter, I asked to see his gun, his hunting license, and his birds. As I was checking his gun for the plug, I noticed that he was keeping a vigilant eye on Belle, who with nose to the ground was working an ever-tightening circle. She located and picked up that last dove I had seen the hunter shoot and brought it over, still fluttering, and dropped it at my feet.

The hunter became increasingly agitated as I counted out the birds from his hunting vest: twelve in all. As I was about to explain to him that he might have a slight problem with that last bird, Belle again began her circular sniffing pattern around the nearby cornstalks. When I saw the man actually begin sweating, I knew that I had a live one on hand. Seconds later, Belle scratched out of the earth three thinly covered and freshly killed doves, all of which she proudly brought over for my examination.

This wasn't the first time Belle had helped me do my job. I once stopped a crabber in Santee Pass behind Capers Island. I was busy checking the baskets of crabs stacked high in his boat to ensure they were of legal size when Belle suddenly went on full alert mode, her nose high in the air. With no prompting from me, she leaped into the crabber's boat and with studied determination muscled her way through a tangle of crab baskets until she was out of sight under the bow. She reappeared only moments later with a freshly shot, still-bleeding "out-of-season" raccoon firmly clenched in her jaw.

Belle was also a good swimmer with a keen sense of direction in swift tidal waters. One night I was working surveillance on the Ashley River opposite the Charleston City Marina. Belle and I were tied up beside one of the darkened craft in the anchorage waiting for a boat suspected of illegal shrimp trawling. We waited, and waited, until finally Belle snorted next to my knee indicating a need for her to tend to some business onshore. I really didn't want to move the boat for fear of being detected. I ignored several more snorts until she punched me in the side with her nose and snorted twice: her "can't wait" signal. I reluctantly eased the boat up on the bank, and Belle disappeared over the bow into the darkness. At that very moment, the darkened silhouette of the suspect boat appeared from behind the marina seawall heading upriver.

I didn't want the culprits to get too far ahead of me. As long as they were heading to their destination, their concentration was forward, but as soon as they put the net over, they would begin looking over their shoulders. I had to act quickly. I called for Belle a few times and when she didn't reappear, I had no choice but to leave without her, stranding her on the island.

I followed the boat unseen some distance up the river and caught them in the act of trawling for shrimp at night in a restricted area. The ticket writing and the seizure of the violator's boat and rigging took hours to complete, and when I came back to get Belle she was nowhere to be found.

With a sinking feeling, I returned to the marina, figuring I would go back and find her in the morning. An old gentleman who frequented the docks was standing by the ramp. He reported that almost an hour before my return he had seen my dog swim up to the bottom of the ramp, shake off, and walk up into the parking lot. In the darkness of night Belle had crossed the entire width of the Ashley River against the tide and had swum around the seawall and through the busy marina. I walked to the far end of the parking lot where I had left my unmarked truck, and she nonchalantly came trotting out from underneath it, wanting a comforting pat.

I've had other Boykins since that old girl's demise, but it's hard to forget her. She rode in my boat perched right up on the bow like the hood ornament on a Jaguar. She exemplified everything I ever expected in a dog. She retrieved everything I ever shot, slept at the foot of the bed with my bride and me, and helped to raise and protect my two girls. On those long patrols she was damned good company.

Chicken Dog

BY CLYDE EDGERTON

In 1971 I began writing songs. The first one was about a dog I'd once owned (lazy 4/4 tempo):

Bird Dog Bell had some
puppies, I think she had ten.
The one I chose was a hotshot–I named him Ben.

The one main dog in my life was Ben, a liver-and-white English pointer, mine over his whole life span, from the time I was sixteen (1960) until I was twenty-six (1970).

My first pet had been a black cat named Inky who one afternoon got run over but not quite killed. I was six, standing in the front yard when it happened. My mama took it upon herself to finish Inky off with a baseball bat (as a good deed). This event marks a milestone in my life. It's the day I learned that when things die, they stay dead. And that sometimes people kill animals on purpose.

Nick and Sam, the two bird dogs my father owned when I was growing up, were family. They somehow seemed more *important* than the pets I owned as a child–Inky, a dog named Sergeant, and a few other dogs. For one thing, they had a job. Hunting. And they had clear human personalities. Nick was cool, distant, proud. Not easy to know. Sam was warm, friendly, the type to walk up, throw an arm around your shoulder, and say, "Hey, man, what's up?"

Early on, I got an idea of how bird dogs should be treated. I still hold onto the rules, prejudices, and guidelines I learned from my father and mother (she was less sentimental) about dogs, especially bird dogs:

1. Mix table scraps and cheap dry dog food with water. Feed your dogs in the evening–after they've been let out of their pen (roomy) to run around awhile.
2. Dogs should have good manners. They must come when called–no hesitations. They should never jump up on a person. (A short section of garden hose to the nose should solve that bothersome problem.)
3. If you don't have time or knowledge to train your dogs yourself, then perhaps your priorities are jumbled.
4. Dogs, especially hunting dogs, should stay out of the house.
5. Dogs should be trained not to chase chickens, for chasing chickens is a serious offense.

We had a few bird dogs after Nick and Sam, and by the time I was sixteen, it was Queen. Her person-like qualities were patience, steadiness, and a kind of mother-hen-ness. She was sired to a dog belonging to a friend of my father's, and on a snowy day, February 29, 1960, ten pups were born to Queen there in the doghouse in the backyard pen. (I changed Queen's name to Bell in the song because I thought it sounded better.) I had first choice and I chose the biggest one, a liver-and-white big-footed prize.

For five long years we hunted many a field and hill, what a thrill.

No, the song didn't win any awards.

During my last two years of high school ('60–'62), my father and I trained Ben. Our methods had been passed down from father to son, uncle to nephew, for generations—among people who hunted in order to put food on the table and money in the bank. (A great-uncle, around the turn of the century, sold undressed quail bound at the neck in Raleigh for a quarter a bunch.)

For puppy Ben—while teaching him to point as Uncle Bob and Daddy showed me—I'd tie a small ball of newspaper onto the end of a string that hung from a fishing pole. Then I'd sling it out in front of him. He would charge it. I'd jerk it into the air before he reached it, and I'd drop it elsewhere. He'd charge it again but perhaps hesitate just a second before getting to it. I'd lift it and fling it out again, and he'd drop into a beautiful point for maybe three seconds; then he'd charge the paper. Within a few days, little Ben, only weeks old, would hold a picture-perfect point for a long time. The instinct to stalk prey and freeze is so strong in some bird dogs that this training is easy, but in others it's not. Ben was easy.

In fact, he was already coming into what would be his eventual person-like qualities: He was smart, enthusiastic, obedient, and *brotherly*.

In the field, Uncle Bob's technique was to keep a long rope around a young dog's neck. When the pup charged a covey pointed by another dog, as he invariably would, Uncle Bob gave him rope, the dog picked up speed, and then the rope gave out just as Uncle Bob jerked on it–a shocking discouragement for running into a covey of quail. Several of these episodes would normally teach a smart dog proper pointing technique. Ben learned everything naturally and easily–honoring (or "backing") another dog's point, retrieving, circling a field with nose to ground, winding, hunting close in thick woods and farther out in thin woods.

Here's the hunt as I knew it in the 1960s: I'd let Ben out of his pen in the backyard, and after he ran around a few minutes, I'd order him into the car trunk and then lower the trunk lid and tie it so it would neither close nor pop open. I'd drive to the woods, a hunting place I knew–or knew about–let Ben out, load my gun, place the safety switch on, and start walking on a general route I knew if I'd hunted that place before. Ben would gallop ahead with his nose lowered.

If it was relatively late in the afternoon, the birds would most likely be feeding in a field of wild lespedeza or perhaps in a patch of soybeans. If I was lucky, I'd come around a bend in a footpath and see Ben pointed in a field, as still as a statue. I'd walk up behind him, talking to him softly: "Whoa, boy. Easy." I'd walk past him, kicking my feet in the grass and weeds. My heart would be pounding. A covey of eight to fifteen quail would explode from the ground in front of me. I'd pick one with my eyes, pull the gun to my shoulder and shoot, swing to another and shoot, and by then the birds would be out of range. I'd watch where they scattered to, as best I could. If one of the two (or in rare cases, both) quail had dropped, I'd call to Ben, "Dead bird, Ben. Dead bird." He'd crisscross in front of me, nose to the ground, find and pick up the dead bird, and bring it to me.

Next, I'd walk toward where the singles scattered into the woods. They would often go for thick cover. "Hunt close, Ben." He'd stay close, and we'd hope to come upon a single or two, or maybe three, with him freezing within a few feet of the bird or birds, so that I could, heart thumping wildly again, kick it up, and shoot. Approaching Ben on a point was intensely satisfying. We'd probably be alone and he would hold perfectly, waiting for me. We worked together with unspoken expectations. After an episode like this, I'd realize anew that I needed no human being with me on a bird hunt.

Ben loved his job. One day I was changing the oil in my car in the backyard. When he saw the open hood, thinking it was hunting time, he jumped up on the engine, realized his mistake, and hopped down.

Around home were neighboring farms with chickens roaming free. When Ben and I crossed a yard or a field near a farm, he'd show an inclination to go for a chicken. After all, what's a chicken but a giant quail that you can catch? The first few times he chased a chicken I beat his ass with a switch, and he learned not to chase chickens. Same with rabbits. After a few training sessions, he'd walk through a yard of chickens and know to stick close to me.

Don't chase those chickens, they ain't
what you're looking for.

I have many Ben hunting stories. Once I shot a bird that, only wounded, ran into a large hole near a big oak tree root. Ben ran to the hole, sniffed, and then disappeared into the opening. I looked in and couldn't see him. I called his name. Silence. Suddenly he emerged with the bird in his mouth. Another time, late in his life, on a Florida hunt with several seasoned dogs of Uncle Bob's, he found eight of the day's twelve coveys. Given the strong talents of the other dogs, this was the equivalent of a couple of grand slams. I was proud of him, often talked about him to friends, girlfriends, anybody with whom I discussed home. Where I was known, Ben was known. I loved to show him off. Uncle Bob would ask, "Where's that barrel-chested handsome bastard?"

When I left home for the air force in 1966, my parents and six-year-old Ben moved to a house without a dog pen. My father, because of the onslaught of emphysema, was unable to hunt much, and Ben became an almost full-fledged family member. Overseas, in the mail, I'd get photo after photo of Ben.

My song, "Bird Dog Ben," ends like this (slow and lazy tempo, still):

I had to go away and stay for
several years.
When I got back, this is what they
told me, I couldn't believe my ears.

(The tempo picks up.)

Bird Dog Ben got lonesome and
started to chase them chickens.

Farmer John got mad,
I'll shoot that dog, he said.
Bird Dog Ben got too close to Farmer
John's back porch.
Farmer John shot him dead.

Then a slower ending with this tag:

A difficult question I'd like
to ask you if I can,
Which is worse–a chicken-chasing
dog, or a bird-dog-shooting man?

I was stationed in Thailand when, in a letter from my parents, I got the news of Ben's death. They didn't tell me who'd shot him, but they let me know there was no argument against the shooting–he'd been shot in somebody's chicken pen. I sort of wanted to know who'd done it, but I never asked. Having recently, in 2009, lost a chicken to a neighbor's dog, I kind of understand. Back then, everybody in the community kind of understood. I'm sure Mama did.

Training Days

BY DAVID DiBENEDETTO

The last dog I owned was a yellow Labrador named Salty. He retrieved one duck in his gundog career. His claim to fame was that after being neutered, he would frequently run away. Where? To the vet's office, a good two miles away. Why? Well, we liked to say he was looking for his balls.

This time around I chose a Boykin spaniel. Or, more correctly, my bride of five months, Jenny, chose a Boykin spaniel. I had never heard of the breed, but Jenny had done her research. The Boykin is the state dog of South Carolina, which was fitting since we had recently moved to Charleston. Boykins love the water and are great around families, and their diminutive stature (the largest ones top out near forty pounds) makes them a good choice for small homes and boats. Best of all, they are bred to be hunting dogs.

But I still caught hell from my circle of hunting buddies. A fellow outdoor writer told me I had gotten "a chick dog," and my own brother told me my pup just wouldn't have the same fire as a Lab.

We named her Pritchard (Pritch for short), after an undeveloped barrier island off the coast. Before she arrived, we had her crate ready, along with an array of toys (including a small beaver without stuffing that we called Roadkill) and a number of bones and rawhides.

We made two trips to the breeder's home before we chose Pritch. We narrowed our choice down to three medium-size females and set them loose, each with a different-colored collar, in the front yard. I tossed puppy-training bumpers to see which dog had the best drive and studied their conformation for indications of character. Then I noticed Jenny holding the blue-collared pup while the other two vied for her attention. Our decision had been made.

When we drove Pritch home, she was a brown ball of fur not more than six pounds. Jenny sat in the backseat because it was safer, and I held the wheel with two hands the entire way. Pritch whimpered herself to sleep after fifteen minutes, and my wife just sat there marveling at our little dog.

Within two days we discovered Pritch had a stomach bug, making the slightest cry reason enough to run to the door with her, hoping to get there in time. We moved her crate from our upstairs bedroom to the living room and inflated the air mattress. There we all slept, "like a pack of wolves," Jenny liked to say. One night I heard Pritch scratch at the crate and, before I could get to her, she let loose. Jenny popped up from the bed.

"What was that?" she asked, as if a limb had fallen on the roof.

"That was your dog," I said, "having an *accident*."

We quickly came up with more words for dog poop than Eskimo have for snow. Pencil. Soft serve. Log. Mud. Before long Jenny and I starting calling our pup Birth Control because there was no time or energy left for nookie. But with a dollop of yogurt and a powdering of probiotics in every meal, Pritch eventually beat the bug, and Jenny and I moved out of the living room—and training began.

I wanted Pritch to be a retriever—a dog that would retrieve the doves and ducks I shot out of the sky. But every gundog must clear three hurdles on her way to the field. She must not be afraid to swim. She must not be gun-shy. And she must be comfortable retrieving real birds.

It's simple, really. Force a pup into water, especially cold water, and chances are she'll never want to go back. Shoot a gun over a pup's head before she's become accustomed to loud noises and she'll run for the hinterlands—every time you fire. And finally, work all you want with rubber bumpers, but if a pup never gets a real bird in her mouth, you can't expect her to pick up a mouthful of feathers on a hunt. In other words, the dog was mine to screw up.

Pritch's first exposure to water did not go according to plan. Jenny and I had driven down to her parents' house in Jacksonville for the weekend. When we got there, I turned on the light in the backyard and let Pritch out. The small pool in the backyard was well lit, but that did not stop Pritch from mistaking it for terra firma. And before I could catch her, she barreled off the pool deck and into the water. I immediately swooped down and fished her out. Back on the deck, she shook like mad and started running in wild circles. That night (and for many that followed) I hardly slept, fearing I had turned my dog against water forever.

But a few weeks later at a small neighborhood pond, I waded in up to my knees with a puppy treat in hand. Pritch was on the bank bucking like a miniature bronco, not sure what to make of the situation. Eventually she stepped in gingerly and began dog-paddling toward me. Jenny was there to catch the event on film, and when I watch the video, I can hear her letting out a gentle whoop.

Gun noise was hard to produce in downtown Charleston, but we did our best. Jenny and I walked around banging pots together like a couple of ragtag musicians without any rhythm. The dog was unfazed. Summer thunderstorms didn't bother her. And when it came time to fire a gun in her vicinity, she didn't flinch.

Introducing her to birds was tougher. Pritch liked her first bird so much she tried to eat it. Jenny was horrified that her sweet little pup had such blood lust, and I feared yet another insurmountable task. A retriever is no good if it eats your doves before you get the opportunity to do so.

In addition to the gundog to-do list, I worked on Pritch's retrieving skills every day at dawn and dusk. A pup's attention span is short. You have about twenty minutes to get your lesson in before your dog decides she'd rather dig a hole than sit, stay, come, or bring a bumper back to you. Between lessons you hope your gundog's life is pretty uneventful. The goal is to make training and retrieving the highlight of the day. But I began to fear this wasn't the case.

Jenny often took Pritch to the dog park, where, in general, mayhem rules and seemingly well-trained puppies pick up bad habits. According to Jenny, Pritch had even made friends with a spastic Dalmatian named Faith and a French bulldog named Lulu. Eventually, I had to quash the dog park excursions, but by then Jenny and Pritch had taken to spending warm days at a friend's pool, the two of them swimming like a pair of debutantes. The final blow came one weekend afternoon when Jenny and I walked into the local Banana Republic and a woman behind the counter looked at Jenny and said, "Where's Pritchy? I have a treat for her."

Eight months after I brought Pritch home, we found ourselves in a dove field on a warm September day. My little brown dog was now thirty-three pounds and had a brand-new blaze-orange collar. My brother, the one who had told me I would suffer for not owning a Lab, volunteered to be the gunner so I could handle Pritch.

Not long into the hunt, my brother hit a dove that fell in some thick sunflowers. I released Pritch, and she tore off through the stalks. As she slowed down short of the fallen bird, I stood up to watch, my heart slowly sinking, but she kept working the area, following her nose. When she found the bird, I blew three short blasts on my whistle as I had done on so many mornings, and she came charging back. "That was pretty impressive," my brother said. "Yep," I said, trying to contain my pride. "My wife knows how to pick a dog."

The Urban Gundog

BY GUY MARTIN

I drove up to get her on a bright September Saturday from the Walking Horse farm in the north part of Limestone County, Alabama. It was a fifty-mile round trip. There was a litter of setters, a sort of good drinking buddy had told me, and then, with the slight wince of apology that's a sign that you're in quail country, he delivered the news that the litter had kennel, but no field, papers. *Irish* setters, he said ominously.

This was what we might call parachute folding. He wanted to be sure he could bail clear of any future bloody accusations, possibly when we were drinking, of outright foppery on his part for foisting such an animal on me.

But here's why we never know when a new dog might come at us: It's simultaneously *never* the right time for a new dog, no matter what, and *always* the right time for a new dog, no matter what. I have learned that it doesn't matter what you think you are doing in your paltry human life. Put another way, the natural world is in alignment when it sends a dog your way. It may be aligned for you or against you.

That said, a new dog was the last thing I needed in this phase. I was driving a badass 1967 two-door Pontiac–rusted through on the trunk–in which I kept having minor wrecks but which, Dorian Gray-style, kept weirdly improving with each wreck. The wrecks were the car's way of letting me renew it piece by piece. I did possess a couple of shotguns and an excellent kit of white tie and tails (both inherited). But I hadn't been hunting in a year, and, no matter how I stared at my calendar, I wasn't being invited to many white-tie cotillions in New Orleans by those staggeringly beautiful black-haired French girls. Bird dog? Worst idea in the history of the world.

But the point is, at the Walking Horse farm, this charming Irish misfit jumped up out of the basket at me. So, there she was. She had good legs. I called her Marlene, after the incendiary Ms. Dietrich.

The funny thing about the dog, which I suspect was also characteristic of her namesake film star, was how many men she instantly put in motion. When I first got her, I lived across the driveway from my grandfather in a couple of rooms on the side of my great-grandfather's house. My grandfather very much did not want the dog tearing up the oak floors or the pine baseboards or the mullions in the old glass front door or anything else that he knew she was gnawing at.

After Marlene survived a couple of his inspection visits–it was, after all, his father's house–Elmer Green's faded blue 1948 GMC pickup rolled up in the driveway with a load of timber in the back that Mr. Green busily set about putting in a corner of the big backyard. Mr. Green was the best carpenter in Athens, Alabama, my hometown. It took him exactly two days in back of my great-grandfather's house to build a pitched-roof doghouse and pen from scratch. He made it with no wasted moves, like God making light.

As I drove back from work on the second day, Mr. Green was stretching the wire between the fence posts and knocking the heavy staples into the posts with his hickory-handled hammer. I let Marlene out the kitchen door into the dusk. She was all of four months old, whirling around our legs like a dervish, bucking with life and busily grinding her nose into us, into the bushes, into the bright new fence, into the very ground.

"'At oughta do it," Mr. Green said. "Hope you're plannin' on takin' her out in the woods. She's got the curiosity for it."

"Yessir," I said. "I'm keen to see what she thinks about things."

"You do that," Mr. Green said to me, his ice-blue eyes lighting up. "You're gonna be in for some fun."

Marlene was great fun. I took along a gun, not to shoot anything, just to show her from the get-go that it was the kind of thing we carried. She stayed way out in front, ranging across the field in a sumptuous repeating arc, nose to the ground. Then she would report back to bump my legs and strike out again. She'd stop and point. I could see the grand computation of the smells running through her brain and then watch her decisions as she worked them out through her body.

Damn, I thought, standing there like an idiot before all this natural sophistication. Where do they get this? She'd never been with any older dogs, but here it was, a giant, gorging, trembling cornucopia of talent. It was too generous, of God, of the cosmos of dogs, of nature. I had no idea where to put the gift.

And I was the wrong man for it. Marlene and I had maybe a half dozen more afternoons in the woods and fields that fall, but a couple of months later, life took a sudden turn and I found myself bound for work in New York City.

My first thought was, who can take the dog? I came up empty-handed: mother, no; father, *truly* no way; and my two brothers were, like me, young bucks on the move. I didn't want to give her outside the family, I think because she'd already become family. However bravely I dressed up the project of taking Marlene to New York City, it was clear that it would be dicey.

Have you ever tried to explain city life to a country dog? It's hard enough to bring *people* from Alabama to New York; a dog poses a different set of problems. The first, main problem is: The dog thinks it's in cotton-town Alabama much, much longer than the people from Alabama can keep that thought going.

Marlene made this clear on our first walk in Central Park. The part of the park we were in has a great bowl-shaped field with a sort of rise on one side and a hollow with some woods and light brush in the back. As her guardian and jailer, I'm not going to go into the tortures of introducing her to the leash–suffice it to say that Marlene jacked that strap to within an inch of its life every time I put it on her.

So, when I took her leash off that first afternoon, she dove headlong into the brush. It was the most fecund, Alabama part of this section of park. I dove in after her. She worked the brush as she would, running back and forth, which meant sometimes I saw her and sometimes I didn't. Then, for a long few minutes, as a block of ice started melting in my gut, I heard nothing.

I'd last seen her quartering away off behind some maples on a west-northwest tangent–Lord only knows what was back there. I'd only been in Central Park a couple of times myself. I ran that way for fifty yards. Nothing. I doubled back, moved east, then started slowly working my way around the compass, as I'd been taught. I was about 270 degrees off my original position when I heard a faint crunching off to the left. I ran that way.

In a small clearing next to a path, just a few feet from an inviting park garbage can, Marlene looked up at me with the pigeon dead in her mouth. There were not many feathers around; her snatch and kill had been admirably stealthy and clean. The problem was that she was now trying to eat the bird–banging the body around in her mouth to get at the breast meat–while rather proudly bringing it for inspection to me. The wings looked funny sticking out of the sides of her mouth.

I bent down and stroked her silky smooth head, then gently took her collar and prized the bird from her. I despise pigeons as much as the next fellow, but there was no way to explain this to

Marlene. To a country dog, a pigeon reads as a dove. And, in Marlene's eyes, this was a really dumb one. All I could do was violate every natural law the dog ever knew: leash her, toss the bird under a bush, and pull her away from the kill.

As the years in the city rolled on, Marlene stopped pointing the pigeons, which did speed up our walks. Still, we achieved some spectacular moments of urban blood sport. One summer day, as we walked down the high-rent part of Fifth Avenue, we came upon a pigeon smack in front of us, pecking at a piece of bread. Marlene instantly lunged; the bird flushed but flew only up to the sill of an open second-floor window. This would be one of those extremely fancy windows framed by a set of silk curtains.

Marlene and I both followed the bird's progress. She glanced at me sidelong, as if to say, *Where's your gun, boss man? I had him for you–why didn't you shoot the damn thing?*

Then we heard a low-high growl, saw a beating flurry of feathers, and the bird was sucked whole into the apartment–in the teeth of the house cat, who had been sunning himself on the sill. It was the best carom shot ever–like falconry, but from the ground up.

Since we were living in New York, of course, it was inevitable that one day we'd be arrested. We'd moved to a loft on the Bowery, at the time a grotty street of restaurant equipment shops and fleabag flophouses. In fact, our loft had been the lobby of one of the flophouses. The new playground was Sara D. Roosevelt Park, a dark, forbidding souk of junkies and prostitutes. Nowadays, it's stupendously fashionable–there's a hotel full of movie stars up the street. But in those years it's fair to say that Marlene and I upped the socioeconomic bracket of the park's citizenry.

As luck would have it, a Department of Sanitation captain with not enough to do that afternoon had been staking out the park for scofflaw dog walkers, a.k.a. me. He gave us a ticket for walking without a leash. To the tune of–we later learned because we rigorously ignored the ticket–$130, plus interest, for the effort it took the city to track us down. The city sheriffs sent us a notice that they were going to come to the house to "collect valuables and property" in the $130 range, which, presumably, the city of New York would then auction off. We never paid. The city finally gave up threatening us.

Marlene stayed with me in New York, living out her field life in aching fits and starts. I tried over the years to get her out into the country–to the islands off Massachusetts, to friends' farms in Vermont–and I made a habit of flying her home with me to Alabama. She was a strong, good flier. It got to the point that she knew Alabama was on the radar when I'd pull the flying cage out of the basement–a sadly limited experience of home, but it was what we could do. After a few years of flying back and forth, I understood what she was telling me: She'd live any kind of life I wanted her to live. It broke my ragged heart that I couldn't give her what she deserved.

As one does with all great dogs gone, I owe Marlene the deepest thanks for all the high and low comedy of our time in the city, and for the grace and forbearance with which she confronted her urban lot in life. For my part, I've one regret: I took her just six times into the real woods and fields of the country in which she was born.

It should have been a thousand.

Dog Tired

BY THOMAS McINTYRE

In our minds, they are forever on the run: hunting, at play, in blurred likeness. What we often fail to recall is that for the better part of each day they are in repose. Which is when our dogs are often at their best.

Dogless for some time since the demise of my last springer, at fifteen, I was for a number of reasons unsure I was ready for the tsunami of a new pup, when I saw my neighbor cutting the grass. It was May; and I knew his liver-and-white bitch—an exceptional hunter—had whelped over a month and a half earlier. I also remembered all the distress that comes with a new dog, from sleep deprivation to property damage; and as I walked up to my neighbor, and he switched off the John Deere, I was almost hoping he would tell me that no pups remained. Instead, he said he still had two, male and female.

He let the spotted balls of blue-eyed dough out of the kennel, and I watched the female bounce across the lawn, then felt a pressure on top of my shoes and looked down to see the male, with the biggest paws I have ever seen on a springer pup, lying on, rather than at, my feet, eyes closed. And perhaps this is where I might make mention of Xenophon.

Son of Gryllus, contemporary of Socrates, a mercenary who fought for the Persian pretender Cyrus, the Athenian Xenophon was the author of *Cynegeticus*. Usually translated as *A Treatise on Hunting with Hounds* or simply *On Hunting*, it is the earliest serious work on the subject. In it he writes, "Hunting and hounds were first an invention of the gods." Along with recommending brusque names, such as Psyché, Porthon, Hybris, Gnomé, and Kainon, for hunting dogs, Xenophon insisted that any good dog must be, among other things, "high-spirited." So it is to Xenophon some twenty-five hundred years ago that, I believe, we owe the usual advice regarding puppies to let the liveliest, most energetic one "choose you." But I digress.

The difficulty was that a lively, high-spirited dog was not what I was in the market for. Let's attribute it to advancing age. Let's say it might have been other considerations. In the language of recovery, I am told, G.O.D. stands for Good Orderly Direction. At this particular juncture of my life, I thought that what I needed was a Good Orderly Dog. So seeing this pup lying, with eyes wide shut, it seemed I had at last found G.O.D.

His eyes did not remain shut for long, though. In short order after I carried him home, he was exhibiting all the pathologies inherent in the springer nature–inordinate affection; unflagging enthusiasm; maddening inquisitiveness; headlong boldness verging on recklessness; approximations of human speech; garbage rummaging and turd eating; a basal inability to lie goddamn still. I wasn't sure how much of this I could take.

I began to hunt Kaycee—named after the Wyoming ranch and town where Nate Champion shot it out with fifty of the Wyoming Stock Growers Association's "Regulators" in April of 1892—when he was five months old. He proved lively and annoying by turns: trying the patience of the older dogs by dangling from their ears until they snarled and snapped at him; dashing among the hunters' feet; picking up a cock bird in the long wet cheatgrass and loping away with it, out of reach, in great adrenalized loops. At night I confined him, sodden and spattered, in his kennel. So I could find some rest.

I'm not sure we entirely appreciate the worth of letting a sleeping dog lie, in close proximity. While some differentiate between "inside" and "outside" dogs, the fundamental ability of a dog to transmit to us its weight and warmth is one of the most elemental aspects of companionship. The defunct "deep ecologist" Paul Shepard offered a scathing appraisal of such companionship, labeling it "neurotic zoophilia." The "great store," he wrote, that we set by such companionship "is because of our modern personal isolation and our sick ecology." He added that there "are still many societies in the world that demonstrate what more fully mature men think of such a companionship . . . by eating the dog whenever they can." To which I say, piffle, sir, sheer piffle. It is the physical presence of a dog—not the concept, and not the nutritional value—that matters to us, in ways we may not be able fully to verbalize. Let alone rationalize. And certainly not psychoanalyze.

So it was hard to take when repeated attempts to persuade Kaycee to sleep peacefully on the bed proved tumultuous. He was all intrusive wet nose, flailing sharp-clawed paws, shredded bedding, and if not outright biting the hand that fed him, then definitely gnawing lightly at it as soon as the lights went out. Which sent him back to the kennel.

Over the many months that followed, I tried calling him onto the bed at night, to see if, finally, he would simmer down. Invariably, such misguided experiments ended in the clanging of the kennel door.

Last October we went to North Dakota. Kaycee was now an adult gundog, his blue eyes a shade of expressive bronze. Miles of walking for me through tall Conservation Reserve Program grass translated into five or six times that much close quartering for Kaycee. Whatever his faults when nominally at rest, there was no denying his enthusiasm for big, wild ring-necked roosters. And for the first time, after a long, hot morning of hunting–high birds haloed in the sun, then falling with feathers drifting behind–I observed something in him I had not detected before, at least to any significant degree: fatigue. He actually sought brush to shade up under, tongue lolling, sides heaving. He drank water by the bucket. Then lay down some more.

That evening in the Best Western, the choice was the kennel box, again, or patting the bedspread beside me. This time, for the first time, Kaycee did something miraculous, for him: He climbed onto the bed and just lay there throughout the night. No thrashing, no nipping, no blanket tugging, just an occasional sigh, followed by faint snoring. And for the first time, with Kaycee, I found the utterly peaceful, reassuring sleep that comes with a one-dog night.

The truth is, a Good Orderly Dog, like a Good Orderly Direction, is a matter of time and patience, which I should have realized. Kaycee sleeps, now, every night, somewhere off toward the end of the blanket, a little out of reach, his presence, when I turn in the darkness, a specific, comforting gravity felt through the covers, like his first pressure on my feet, aiding and abetting my repose. Call it what you will, but for me it is the most substantial quality he possesses, the best part of him and the best part of the day. Or night.

Doubling Down

BY JEFF HULL

Juice was black. Sola was white. In many ways it was that simple. Juice came first. I was living at the time with a lovely, lively woman from Alabama and we were toying with the idea of making our situation . . . less escapable. Well, she was. Raising a Lab, I told her in some desperate fit of placation, would be excellent practice for raising a child.

Juice's father was a show dog, his mother descended from field trial champions. He was, even in infancy, a tremendously handsome dog, and would grow into an animal packed with sleek muscle, classically boxy of head and burly of shoulder. He was, too, a bundle of exuberance, never bored, curious about every corner of his world.

Until I got Sola, I thought Juice was a smart dog, and that I was a clever trainer. I trained him straight from the pages of the classic Richard A. Wolters book *Water Dog*. On top of that, when I lay on the couch at night watching movies with an enormous bowl of popcorn coated in butter, Cajun spices, crushed garlic, turmeric, and Parmesan cheese, Juice sat on his haunches facing me, stalactites of drool drooping from his muzzle. I pointed to my right, then threw a piece of popcorn that way. I signaled left, threw. This, I thought, would teach him that I knew where things fell.

But Juice tended to live by instinct. He was intimately acquainted with trouble, and he seemed comfortable with the relationship. Juice was the kind of dog that, given a sufficient window of inattention, would swipe plated food from the counter and then attempt to salvage tidbits from among the shards of broken plate before the trouble came down.

Still, I loved the big hunk like crazy.

Enter Sola. I decided to add another dog to my life long after the Alabama girl had departed for points . . . more promising. Juice, I thought, was lonely and would behave better with a dog friend. I have no idea why I thought this, but I decided it all lying in a sheep pasture in Iceland reading Halldór Laxness's *Independent People*. The name of the primary female character in that book was Sola, an Icelandic diminutive meaning "Little Sunshine." Within a week after returning to the States, I drove up to a breeder's house and I saw, in the middle of the dirt driveway, a pure white butter-belly Lab pup, sitting on one hip, ears perked. Sola.

Technically she was a yellow Lab, but only in the saffron-stained tips of her ears, her black nose, and her deep brown eyes was she anything but white as the driven snow. My notion of having two dogs who played and ran together, exhausting themselves and their need for my attention, was daffy. Juice was already six years old and heavily dedicated to sleeping. He tolerated the new pup but did not volunteer to be a mentor.

And Sola was not much interested in other dogs; she decided early on she was one of the People. She retrieved tennis balls, but not with the obsessive zeal of Juice. She took herself swimming, but not for the hours of biting at splashes Juice enjoyed. Soon enough, I would understand why. But until then, I saw Sola as a sweet, gentle soul, always curled next to me or partially on me, her efforts and energies dedicated to loving me.

By the time she was two, I knew something was wrong with Sola. Her hips, my vet told me, were shot. Dysplasia. I had a little money then, so I gave her a titanium hip.

Was Sola so much smarter than Juice, or was she simply blessed with the capacity to understand what did and didn't work out so well for the black dog? Or to care? What I know was that Juice slurped up piles of goose shit by the pond, often glancing up at my exhortations for him to cease only long enough to gauge how fast he had to move to swallow a few more piles before I reached him. Sola stopped with a simple "Nope." Juice clambered to the front seat from the back to swipe sandwiches during gas stops. He regularly raided unguarded grocery bags. I could place a piece of steak scrap in front of Sola's nose and she'd wait until I said "Okay" before digging in.

Juice loved to hunt, yelping and bellowing like a dog being skinned alive if I tried to carry a shotgun out of the house and didn't take him. Sola approached hunting like everything else—she was good at it, she knew that. She was not unenthusiastic. But if I wanted to stay home, lie on the couch, and rub her ears, that would have been fine.

We turned to pheasant and grouse and partridge, and Sola learned from Juice—things not to do, primarily. Juice bulldozed through the brush, his exact location determinable only by watching the birds he busted into the sky, far in the distance. Sola gathered how frustrated that made me and circled back. The first time she worked her way back toward me and pinned a bird down, it surprised her; then she adopted it as strategy. Juice always seemed baffled and put out that I couldn't shoot a bird he put up six hundred yards in front of me. As a fellow People, Sola understood my role in the game and my limitations. On more than one occasion, she stood, ears cocked and head teeter-tottering while she regarded a spot in front of her, waiting for me to bring the gun—and, as I approached and the bird flushed, leaped into the air and caught it in her teeth.

But she also learned from Juice's tenacity. On one of her first hunts, Sola quit early on a wounded bird, coming back to me for direction. Juice did a horizon job, disappearing over the curve of the earth—I caught a glimpse of him a quarter mile away, cresting a roll in the land. But twenty minutes later he was back, bird in mouth. After that, when the gun went off, Sola followed him unless I redirected her.

Juice, the dog acquired to stave off certain family-making urges, died at the age of sixteen and a half, nine days after my first son was born. He was shot through with cancer and should have gone earlier, but the birth of my boy befuddled me and I wanted to hold on to this final vestige of my former life. I dawdled to the dog's detriment. When finally Juice's suffering broke through to me, I arranged for a vet to come to my house, then took Juice for a last swim in our pond. I brought his sleeping pad to the pond's bank, let him lie comfortably in the sun, and tossed him popcorn kernels soaked in butter and cheese from a huge bowl. When the vet strode across the lawn, Sola rose and moved away, watching nervously from some remove even though the popcorn was still flying. The first needle went in his leg and Juice tried to get up, fighting me to break free. I remember carrying his dead body across the yard to the vet's pickup, how unbelievably heavy he felt.

Sola was an only dog for three years. She had grown fat due to lack of exercise–her bad hips came bundled with a fused spine and arthritic hocks, so that even swimming caused her pain–and I thought her precipitous weight loss and lack of energy were a result of a severe diet I'd put her on. It was liver cancer. I made the call the day I found out. That afternoon the same vet came to my house. Sola had her popcorn on the lawn but was too sick to eat more than a few kernels. She went so quietly, the way she had lived, her head on my lap, my hands stroking her ears.

Carrying her along the same path on which I had earlier borne Juice, everything about her seemed so light.

The Sweetest Sound

BY RICK BRAGG

I cannot remember everything, but I remember that stingy moon, remember a sliver of cold in a black sky, thin and useless. You might as well try to light your way with a piece of broken glass, for all the good it was. My flashlight, a relic from the Korean War, had died about eighteen seconds after full dark. The boys in front of me, friends, brothers, and cousins, mostly, were barely better off. They had to shake their hand-me-down batteries every few steps to coerce a last feeble glint of electricity, but I could have shaken mine like a Birmingham hoochie koo and still been walking in the dark. Only my brother Sam, who was born grown, who already had sideburns at age thirteen, had a good battery in his light. He was irritating that way. People said the day he was born he just dusted himself off in the hospital and walked home.

But, except for the threat of falling to our deaths down some crevasse, I guess we did not really need light. Our direction was determined not by what we could see but by what we could hear. We followed the sound across that frozen ground, followed a thing faint, constant, lovely, as we trudged single file across the haunted ridges and deep into the black hollows, the mist shining silver where that one good flashlight bored through the dark.

I can still hear it, after all these years. The poet in me—well, the never-was poet in me—would like to call it music, but it was prettier than that, prettier than I can say. It would have been nice to sit on a porch and hear it, but the other boys looked at me funny when I suggested that. Only the boys afraid of the dark or the cold or monsters stayed home. I went, this one time, to show I was not one of them.

I was the youngest, and so the last in line. As we made our way diagonally up a ridge, rocks turning our ankles beneath the slick carpet of leaves, I felt myself begin to slide sickeningly straight down the mountain, straight toward what I knew to be a bone-breaking deadfall. I caught myself on a gummy pine sapling, breathed a minute, and started up again, even farther behind. No one had even turned around, and the romantic in me, the one who read about lost souls on desert islands, wondered how long I would have lain there, broken and forgotten. My brothers said I thought like that because I read too many books.

But falling was not romantic on the mountain. Falling was what you did up here. You walked; you fell. You chewed some Brown's Mule, or some Beech-Nut, if your stomach could handle it. I did not chew, so mostly I just walked and fell.

What a dumbass I was, I thought, as I slid again and lost thirty yards of the uphill ground I had gained. A smart boy would have chased some brighter light, somewhere, because the light was where the girls lived. A smart boy would have been in town, leaning on the hood of a car at the Rocket Drive Inn with a Cherry Coke in one hand and a beautiful woman in the other. Or, at least, that was how I figured it should be. I was not yet ten years old, and a beautiful woman would have sent me into a convulsion.

No, we went the other way, away from perfume and soft shoulders, gouging deeper and deeper into the dark, into the foothills of the Appalachians along the Alabama-Georgia line. It was November, maybe even as late as December, 1969, but it could have been any night when boredom was stronger than common sense, after the cold sent the snakes down into the earth, and walking out into nothing was as much adventure as we could divine.

My big brother, Sam, was only three years older than me, but he drove a Willys Jeep he had hacked out of a rust pile and made to live again by soaking its bones in buckets of dirty gasoline. And so, he got to walk in front. He had an ax in a tow sack, but no gun. This was as far from a gentleman's hunt as I guess a fellow could get.

We were all the same, us boys, on the outside. We did not own big parkas or camouflage anything, because while we still had real winter back then, it was too short in which to invest much wealth. We wore flannel shirts and thermal undershirts and something called a car coat, a thin and useless thing that, as near as I could tell, was made out of polyester, cat hair, and itch. Walmart would, one day, sell a trillion of them. We got a new one every other year; that, and a gross of underwear.

The smart ones in the group wore two pairs of pants, even three, because the briars ripped at our legs with every step. Sometime, back in the times of our grandfathers, these mountains had been old-growth hardwoods and towering pines, but none of us could remember a time when the South looked like that. Old men talked of an age when the great trees towered into the clouds and the forest floor was dark and smooth and clean, but these mountains had been clear-cut generations before, creating a tangled mess of skinny trees fighting for the light, with undergrowth and saw briars strung between them like razor wire.

It was a time before hunting was a fashion. We hunted in our work boots, laced up around two pairs of socks–three, if you were growing into them. The ones who had gloves wore them and the ones who didn't walked through the woods with a pair of tube socks over our hands. I guess an outsider would have laughed at us, but outsiders did not get to go.

So armored, spitting, and breathing hard, we attacked the mountain. And no one said a word. I tried to whine, once, and ducked just in time to avoid being slapped back down the mountain.

"Hush," my brother hissed, then, gentler: "Listen."

The baying was so thin it vanished in the wind in the trees.

But he could hear it plain.

"Joe," he said.

We often got things, back then, no one else wanted. We were the poorest relations and naturally became the repository for things cracked, busted, rusted, or slightly burned. Kinfolks and other well-meaning people brought us these things with a straight face, but we knew. The dump charged between three and five dollars a load, while we took their castoffs for free, took their refrigerators that did not cool, and fans that did not spin, and big, heavy televisions stuck permanently on a horizontal roll. "You can fix it," they always said, and drove away. At one point, my grandmother Ava had three radios on her dresser that were as mute as a stone. People even brought us tires worn completely through. "You can get them recapped," they said. Now how in the hell do you do that, unless your last name is Goodyear?

That is how we got Joe.

Joe was a coon dog, a fierce mixed-breed dog, a mass of quivering muscle and intelligence. You could see the black and tan in him and bluetick and even some red feist, but mostly he seemed composed of thin white lines where his body had been ripped and torn, as if instead of breeding those bloodlines into him some Doctor Frankenstein had just sewn him together from spare parts. He had a savaged snout and no ears, none. Coons had chewed them off at the skull.

We got him not because he was wounded or finished, but because he would not stay in a pen. He belonged to a tough old pulpwooder named Hoyt Cochran, who penned him with a gyp named Bell. But Joe was a climber, and so, when no one was looking, he would scale the chain link or dog wire like a monkey and go hunt free. He did not run trash and he was fearless and he would hold a tree for five hours. "Hold a tree all night," my brother said. But a dog that cannot be kept is a worrisome thing, so Sam bought Joe cheap, probably saving him from a bullet.

He drove a steel rod in the dirt of the backyard and fixed to that several feet of heavy chain. Joe lived most of his long life that way, staked down between his water bucket, food bowl—really an upside-down hubcap—and his doghouse, a plywood shell covered over with shingles and filled with fresh hay.

I would watch him there many days, my heart breaking a little, because nothing should live like that. Every now and then the wind would carry a scent to him there and he would drag his chain in the direction of that smell and pull it tight, his ruined nose twitching, and bay.

It came to me eventually that his time on the chain was not living, and the only time he was really alive was on the mountain. There, he was something different. "He beat a lot of dogs, a lot of expensive, purebred dogs," Sam said, years later. To him, Joe was just one more discarded thing he was able to get some good out of, his way of saying to the discarders, yeah, the joke's on you.

Joe knew it, too, knew that he was something more. If he treed a coon with a pack of lesser dogs, or just one noisy dog who barked if he treed a coon or barked if he scratched himself, he would set himself apart, and fix his head in the direction of the coon and hold it there, as if to say to the humans, If you were gonna bring along this white trash, what did you need me for?

"He didn't like an ill dog," Sam said. "He meant for the others to behave."

He hunted him with a purebred black and tan named Blackie, a gift from an uncle. Blackie had a lovely voice but was prone to get lost and wander for days.

But together, off in the distance, they made that beautiful sound.

Coon dogs have one bark when they strike a trail, an excited, insistent one, and another, steadier, melodic, as they trail, and a third, urgent, excited, when they tree. But it was the trail bark that was most lovely.

"Joe had a kind of sharp 'yerp' sound, and Blackie had more of a 'yowl,' and when they were on that trail that sound would get to rolling, kinda, in the distance, and that was what was so pretty," my brother said.

YERP!
YOWL!
YERP!
YOWL!

And the boys in the cold and the dark forgot about falling off the mountain and slid and scrambled in the direction of that sound, faster and faster. It was the pure excitement of it that hurried them, not the chance that Joe would not stay treed. Sam knew that when he got there, there would be a coon either in the tree or dead on the ground, because if Joe caught one in the open, it was done. "He learned to kill them without getting hurt real bad, but not in time to save his ears."

This night, we were still what seemed a mile away when we heard that change—heard the baying ratchet itself up—and knew that the dogs were treed.

"Talk to 'im, son," my brother hollered, and the other little boys whooped, and I swear one or two of them jumped for joy. To say we didn't get out much does not nearly cover it.

We got there in time to see Joe running a wide circle around the tree, once, twice. That way he made sure that the coon had not outsmarted him and slipped away. Then he put his front paws on the trunk, and sang to us. Our one good light played up into the branches of a pine, and there he was, a big boar coon. I know that a big boar was more than enough for most dogs, all teeth and razor blades. But he did not look big or dangerous. He just looked scared, behind his black mask, a thing I knew to be deceptive. Sam climbed the tree and, carefully, with a small ax, knocked the coon to the ground, and the dogs closed in. That might not be the way others did it, the purists of this sport, but it was the way a troop of bloodthirsty redneck boys did it in the winter of 1969. I do not remember a terrible fight, because Joe was such an assassin and was on his throat in an instant, but I remember an odd quiet when it was done, and boys, looking like little boys again, milling around, not sure what to do with their hands. We put the limp coon in the tow sack. There was a man in Jacksonville who paid cash money for it, not for the skin but the meat.

My brother hunted another forty years, and still does, when his wore-out body will allow. He grew gentler—not a lot, but some—and sometimes just hunted to hear that sound, and dragged his dogs off the tree and went home once he knew the pretty part of it was done. He saw the sport change, and changed with it. He hunted on into his fifties, on into an age of tracking devices and shock collars. He carried a GPS. He asked me once to put the shock collar on so he could test it, but I have gotten some smarter since 1969.

But my time in the mountains, listening to the dogs, ended when I was a boy. When I think of it, which I often do, I am always ten years old, waddling across that mountain in so many pants my legs won't bend at the knee, knowing that if I fell I would never get up again.

Joe lived to be more than fifteen years old. "I finally just turned him loose," said my brother. "His teeth were wore down to nothing. But you know, I'd come home from work and he'd be gone, and I'd hear him off in the distance, and I'd have to go and get him off a tree, somewhere behind the house." It may be that there was a coon up there, or, near the end, it may be that he just thought there was. I guess there is no reason to think he would be any different from people, in that way.

Every now and then my big brother will ask me if I want to go for one last walk up the mountain, behind a new dog, but I just say, no, I'll listen from here. From the porch of my mother's house you can sometimes hear that sound, faintly, drifting down through the trees, and I shift my weight on the boards of the porch, and think how fine it is that an old man does not have to prove he is any kind of man, anymore.

CHAPTER 3

Man's Best Friend

Lapdog

BY CHARLES GAINES

When I was a child I had polio, and the doctors thought I would die. Then one day my mother went out and, with the last of her meager savings, bought a Yorkshire terrier. My father, who had recently lost his job at the factory, named her Trixie, and every day he would put the little dog on top of my iron lung so I could watch her. Somehow Trixie awakened in my frail breast a will to live, and when I finally recovered I determined that one day, God willing, I would have a dog just like her.

Well, not really. But that was one of the stories I used to offer up to my bird-hunting, waterfowling, coon-hunting, rabbit-hunting, fox-hunting friends when asked how it was exactly that I had come to own a Yorkie. Another had to do with how a Yorkie had saved my platoon and me in Vietnam by sniffing out a tunnel full of Vietcong. Ah, the perils and temptations of false pride! But more on that later.

The *true* story is that thirteen years ago, my wife, Patricia, and I took a canoe-camping trip in North Carolina with our daughter Greta and her Yorkie, for which Patricia fell hard. Two weeks later Greta sent us a female puppy from a trailer park kennel in Tennessee. Patricia named this scruff muffin Dixie Belle Berubi (the kennel owner's last name) Ellisor (Patricia's maiden name) Gaines. For nearly a decade and a half she has gone by Belle to her friends–and they are legion.

Now, dogs and I go way back. I grew up in a house full of them, and Patricia and I have owned fifteen. But before Belle they were all males, and all but two were either upland-bird or waterfowl dogs–working dogs–dogs definitely bigger than a dust mop. Belle was something distinctly new for me; and something, I'm chagrined now to admit, I was a tiny bit ashamed of. As well as worried about.

When Belle came to live with us, we were in New Hampshire, a fairly benign environment for a lapdog. But we were about to drive up to Canada, where we live for part of each year on three hundred wild acres that are home to a large and hungry population of coyotes. "She won't last two weeks," I told Patricia with a mixture of scorn and trepidation (the latter because Belle had already made my lap her favorite nest whenever I was reading or writing, and I had come to find her presence there . . . not altogether unamiable).

To give her at least a chance with the coyotes, I decided to train her to come whenever I blew a whistle. I had trained ten hardheaded male dogs to do this, so, I figured, how hard could it be with this little thing? Unquestionably too hard for me. And also for my hunting pal, C. V. Child, who is known for a Teutonic hand with dogs. On his annual woodcock and grouse trip to Nova Scotia during Belle's first autumn there, C. V. undertook to train her to come and heel, and worked at it slavishly for days with no more luck than I had had. On the command to "come," she would do so if a treat was in the offing or it suited her whim; otherwise, she would look at C. V. and me with amused disdain, like a debutante whistled at by construction workers, toss her head, and trot off about her business.

Increasingly, that business became rabbit-hunting forays into the woods and wild rose thickets of our property that would often have her missing for hours, during which Patricia and I could do nothing more than chew our nails. Once, with night falling, we drove out in the woods into which I had watched Belle disappear around noon, hollering hopelessly for her to come, and left our truck there with the doors open, the headlights on, the engine running, and Dolly Parton singing at top volume from a tape. Maybe it was the call of her hillbilly Tennessee genes, but an hour later we found her curled up in the driver's seat snoozing to "Coat of Many Colors."

For twelve years since then, Belle has demonstrated over and over that her staying power is at least equal, and perhaps related, to her stubbornness. She has insouciantly eluded coyotes and bald eagles in Nova Scotia; survived many daylong and a couple of overnight jaunts in the Alabama woods, where shooting dogs is considered a sport; picked up in her mouth and killed a water moccasin; and outlived eight of her bird dog brothers–all while doing absolutely nothing she is instructed by human beings to do for her safety.

Which raises this question: Is Belle, in fact, a "good dog"? Well, certainly she is always willful and often naughty. And it must be admitted that she is no Westminster beauty queen with a long, silky coat and come-hither eyes. Belle's eyes are mostly defiant except during thunderstorms, and she is happiest when her coat is clotted with tangles and bits of briar and leaf. She is also temperamental, impatient to the point of disgust with all male dogs, and haughty to other females. And she barks peremptorily for table scraps.

All these things I, and she, will readily admit to. But also these: No dog of the dozens I have slept with is better in bed. By that, let me be quick to add, I mean that she will cuddle against your stomach or back like a hairy, heated neck pillow for twelve hours straight if that is your wont (as it often is Patricia's), adjusting perfectly to your position with a fetching, soporific little snore. And perhaps to please me, but more likely not, she has become a good if improbable sporting dog, flushing and treeing ruffed grouse in Nova Scotia and fetching the bluegills and bass I catch from the banks of the lake we live on in Alabama. Unquestionably, she has, in face-card spades, that most clearly defining and touching of canine characteristics–loyalty–as well as a deeply feminine intelligence, a bighearted capacity for spontaneous delight, and infallible taste in people. And there is this: She grins. Do all good dogs grin? Perhaps not, but they should. Finally, did I mention that she holds her paw up to her mouth when she grins? Well, not really, but I am trying to train her to do it.

Good dog or not, somehow this Yorkie over the years has become talismanic to Patricia and me; and, to our friends, emblematic of us and our life. For the past thirteen years, every visitor to our home in Nova Scotia (if it is a visitor we want back) has left a signature in a guest book alongside a photo, taken by my wife, of him or her holding not one of my glamorous setters or golden retrievers, or the photogenic English cocker Sumo, but Dixie Belle Berubi Ellisor Gaines.

Back to the aforementioned false pride: On the day I finally shed it, some two years into Belle's reign in our household, I filled my lower lip with Skoal Long Cut and drove with her in my pickup down to the lobster dock in East Tracadie, Nova Scotia. Reggie Beshong, Angus Cotie, and the other fishermen had hauled their pots for the day and were sitting around the dock having a smoke. I drove up to them, rolled down the window, propped my Yorkie in the crook of my arm so she and I were looking at them face by face, and said, "Howdy, boys. How was the fishing?"

Well, not really.

King of Oxford

BY JIM DEES

My buddy Ronzo once gave away a litter of puppies only to discover some years later that one of them was being kept in an electrified pen. The new owner, a farmer friend, told him the dog had been "gettin' after" his cows. Ronzo negotiated the return of the thoroughly freaked-out, now four-year-old black Lab, whom the friend had named King. He turned out to be a wounded soul with untrusting eyes and a prematurely white snout. Ron Shapiro–Ronzo to his friends–and I were roommates and business partners back in the mid-1980s. We lived in a hippie cabin seven miles in the woods outside of Oxford, Mississippi. Those years were our '60s. Oxford was just beginning its evolution from hickville to litville.

We operated the Hoka, a bohemian coffeehouse/movie theater, named after the Chickasaw princess who had once owned the land that would become Oxford. The converted cotton warehouse catered to Ole Miss kids and a variety of oddballs with a menu of art films, wheat bread sandwiches, and killer desserts. Dogs just happened to be allowed (we bought off the health inspector with cheesecake).

King, as might be imagined, was skittish at first. For months he wouldn't come out of the Hoka's back office. Little by little he emerged, the aromas from the kitchen easing the way. Finally he found a favorite chair. He even started watching movies in the theater.

This "dogs welcome" policy certainly found favor with the locals, including the late great writer Willie Morris. Willie had recently rebounded to his home state after a divorce and a heroic but controversial tenure as the youngest editor of *Harper's Magazine*. He spent the 1980s in Oxford, and he and his black Lab, Pete, were Hoka regulars. Pete would coil up at Willie's feet as the author downed bourbon-laced coffee at the counter, usually closing the place at 2:00 a.m. and moving the party to his nearby faculty bungalow. Pete shows up in many of Willie's pieces and on the cover of his brilliant collection *Terrains of the Heart and Other Essays on Home*. The cover photo shows Willie ambling across the field of the Ole Miss football stadium, empty on a quiet weekday–alone, windswept, forlorn. Behind him a few paces, if you squint, there is Pete, shadowing Willie like a happy ghost. When Pete died, Oxford grieved, and he was actually interred in one of the town's human cemeteries, named, of course, St. Peter's.

The loss of Pete left Willie bereft, to use one of his favorite words. He began asking Ronzo if King could visit, and these cheer-up soirees invariably turned into sleepovers. Eventually we wouldn't see King for two or three days at a time.

King had blossomed into something of a local celebrity himself. He and Ronzo were dog and man about town, showing up at most any event or just presiding at the Hoka. Most memorably, they could be seen tooling around the Square in Ronzo's massive white Oldsmobile convertible. The vehicle was a distinctive land yacht among the college-town Beamers and Jeeps. It lumbered along like a parade float with the top down and Ronzo at the wheel, his shoulder-length hair waving at passersby. King sat in the backseat, well, regally, chauffeured by a human.

One day Ronzo and I were tossing a Frisbee before a long Hoka shift, and Ronzo sailed one way over my head and it kept sailing. King, without prodding, charged after it. Leaving the ground at the precise moment and fully extending himself, he snared the disk in his teeth before flying to a stop and casually returning it to Ronzo. We rejoiced in this. And once Willie witnessed King's feats, he persuaded a fellow dog lover, then Ole Miss head football coach Billy "Dog" Brewer, to have King perform at halftime of the

Vanderbilt game. King missed the first couple of throws that day, from an unfamiliar cheerleader. But he soon found his rhythm, and by the end, the entire stadium was roaring. Ronzo and I were standing down on the field, on the sidelines, and had the rare thrill of feeling forty thousand voices in full throat cascading from above. We later joked it was the most offense the crowd would see all day.

King even rescued *me* late one night during a war games party thrown by our good friends Ron and Becky Feder. At a certain point, late in the evening, Ron put on side three of the *Apocalypse Now* sound track (the original vinyl), which consisted mostly of manic machine gun fire and chopping helicopter blades. He cranked it. Whoever didn't leave, he divided into teams. He equipped each team with Roman candles and flares, and we hiked two short blocks down to Rowan Oak, William Faulkner's house. Surrounding the house was (and still is) a pleasant forested area, Bailey's Woods, with a walking trail and a wooden footbridge spanning a dry creek bed. During the fog of war, I stumbled around a deep thicket and sustained a gash on my left leg. It stung as only being stupid can. Thick sheets of the gunpowder smoke enveloped the woods, and I suddenly had no idea how to get back to my team. I could make out a weird shadow shifting near the ground ahead of me, so I moved toward it. The shadow turned out to be King, who calmly led me back to the main path. I could see him only because he contrasted with the darkness. He was blacker than the night, and I'll always be grateful for his grace under craziness.

Willie's dog grief became exacerbated as the long Pete-less nights wore on, amplified by bourbon and Viceroy filter tips.

During one liquid late night, with Ronzo and me out of town, Willie paid his tab and told the Hoka kitchen staff he was taking King for the night. Willie and a small group, including an overbearing New York transplant, Terry B., piled into Willie's Buick, with King riding shotgun. At Willie's house, after several more rounds of Jack Daniel's, Willie and Terry B. decided it was time for convenience-store chicken legs. Terry B. grabbed Willie's keys and said he'd be right back. Unbeknownst to anyone, King slipped out and dashed after the car, galloping down the middle of the street. He was clipped by an oncoming vehicle that never stopped. Terry B. said later, "I heard a sound that was wrong." He pulled over to check, and King was dead in the road. In this pre-cell-phone era, Terry B. was left to hysterically scoop up King by himself and figure out how to break the news to Willie, and then of course, ultimately, to Ronzo. Causing the death of your friend's dog, even accidentally, is an adult dose of anguish, an electric pen around your brain.

We buried King in the backyard of our cabin. Willie came out and spoke a few tortured words. Dog man that he was, he never got over it. Terry B.'s heart-searing guilt was so severe, he left town. Ronzo and I went through all the stages of grief, most especially anger. But anger isn't how we remember the dogs of our lives. The anger is no match for the bright memories–in our case, of a flying dog, rescued from fear and bringing thunderous cheers from an admiring throng. And then going for a spin with the top down.

Traveling Companion

BY DONOVAN WEBSTER

He came to me as a gift.

In late August or early September 1997, I mentioned to my wife that I'd love "a big American-style chocolate Lab that I can take hunting. . . . You know, a dog that won't quit until the day is over."

Then, not long after, I left for a magazine assignment in the Sahara and forgot all about it. When I got home, the day before Thanksgiving, Janet and the children met me in the airport's baggage area. Everyone hugged.

Then my youngest–Anna–all of two years old and white blond, couldn't contain herself any longer. She was hugging my leg. I was exhausted and waiting for the baggage, just wanting to go home.

"Daddy! Daddy!" she shouted in the airport as she shook me. "We got you the dog you *wanted! For Christmas!*"

....

They had already named him: Traveler. My son had generated the name, not for Robert E. Lee's horse, but because James wanted the dog to travel with us. When we got to the car in the airport lot, the puppy was in the backseat. He was an eight-week-old mound of sleeping brown fur, who–I saw when we woke him up–had hazel eyes that looked at everyone like they were his best friend. We all said hello to the puppy.

And then the adventures began.

He was a high-energy maniac almost immediately. At least, once he got his bearings around the house and property. He'd tear off after birds, squirrels, rabbits, foxes–pretty much anything that moved–and when he wasn't doing that, he'd be nipping at your trouser hems or going after your feet (at one point, he even chewed up a pillar to the right of the house's front door). Sometimes, standing in the kitchen, he'd take a run at me, leap as he got close, then grab a loose fold of my blue jeans and swing there until the pants ripped. With his needle-like puppy teeth, he cut a swath through my entire family's clothing, from shoes to socks to anything he could get close to.

But he was also growing into an incredibly beautiful dog: heavily muscled, and with a sheeny brown short-haired coat. He could also be extremely sweet. As I work at home, he became very devoted to me (it's probably fair to admit I was the love of his life). And while he may not have won any MacArthur grants, he did learn how to hunt or find any human food that might have been dropped on the floor. Eventually, as he cleaned up a hot-dog spill one evening from one of the kids, my wife gave him the nickname FloorMaster 2000.

He also grew up fast. When he was four months, we began early hunting training. You know: sit, stay, down, go, get it, come, leave it. He learned fairly quickly but was also willful until he felt like playing along. And he continued leanly muscling up. When he was six months—and sixty pounds—we introduced him to water. At a year, he began to go with me on my daily five-mile runs at lunch, and was always good to—as he'd been trained to do—return to heel when the noise of a car or truck became audible.

On the runs, down the country roads behind our house here in Charlottesville, Virginia, he would see a deer off in some field, and tear away, thinking he could catch it. Most days, while I ran five miles, Traveler probably logged closer to ten or fifteen. As he ran, you could see the muscles over his ribs move in the sun. He was now somewhere between eighty and ninety pounds.

By then, beyond the Charlottesville place, we'd secured a large two-room cabin on a bunch of land surrounded by more than 100,000 acres of national forest about an hour west of our house. At the cabin, Traveler roared around the landscape all day, constantly running over several square miles of land, and then would come home in the late afternoon and fall out with fatigue on the front porch. Sometimes, as the kids played in the river with friends, I'd go out and stand in the river, throwing him a tennis ball, which he would always return, swimming upstream, to me. Then, I'd just stand there, with the water too deep for Traveler to stand in, so he had to swim to stay at my side.

At the cabin in the evenings, after checking the property and now away from the river, Traveler was often at my feet as my wife finished up baking a chicken or cooking some pasta with homemade sauce. There'd be 1940s jazz on the CD player; the kids would be tired from playing in the river all day. And at the end of the days at the cabin, Traveler and I talked a lot as we sat on the porch, watching the day become night. To be honest, it was a one-sided conversation as I sat in a porch chair and he lay on my feet.

But come to think of it, maybe it wasn't such a one-sided conversation. After all, he was saying he was there for me.

But mainly he went with us to places like Chicago, D.C., and Memphis, where Traveler would stay at my heel on sidewalks and in restaurants, without need of a leash.

In those situations, my daughter would always say: "Traveler has his city manners on." People would compliment him on his good behavior. At the Topaz hotel in D.C., he was a welcome and beloved guest. And when I had meetings that ran late, I could call the front desk, and someone would go to my room and take him out for a walk. For that, we came prepared with a leash, but he didn't really need it.

He went with me everywhere: to lunch meetings and on errands and on car trips as I did reporting. He loved my in-laws' lake house in Arkansas and my family's summer house overlooking Lake Michigan. He loved to swim. And swimming with him was fun, if sometimes dangerous, as Traveler upon occasion thought you might need saving so he would come over to help. The problem was that if you went underwater to escape his assistance, he, too, would go under with his paws and mouth all over you in order to save you. It left scratches.

But, hey: It was done out of love.

Over time, my old F-150 smelled of wet dog and spilled coffee any time the day got warm. On any trip, Traveler always dozed on the backseat of the crew cab, patiently waiting as I went into a store. He just loved being out, in the truck, with the windows down. And, while he never actually said it, I think he loved being out with me. He had become my shadow.

Finally, in early 2010, Traveler's health began to fail. Time was taking its toll.

Now he didn't go for runs with me every day. Sometimes, as I got out my running shoes–an act that used to make him begin panting and pacing in expectation–he'd look over from his bed, acknowledge what was happening, and go back to dozing.

He had become gray faced, and increasingly he had started to fall down involuntarily and become incontinent. When that happened, Janet and I would come over and comfort and stroke him until he was strong enough to stand again, then we cleaned up what had happened. At the Topaz hotel, when I didn't bring him on one trip, the person at the front desk asked: "Where is Traveler?"

He died at my feet on a January morning. The veterinarian, who had come to officiate, and I drank coffee first. Traveler didn't move. I apologized to Traveler for not going hunting with him more regularly, stroking his cheek as I said it. We took him outside. He died while looking for birds in the sky on a beautiful, cloudless, white-snowy day.

When it was over, both the doctor and I cried like babies.

There are and have been other dogs in the house, but I still miss Traveler every day. He lived up to everything I wanted, and never quit until any day was over. It turns out, somewhere along our thirteen-year journey together, Traveler had become my best friend.

As I said, he came to me as a gift.

A Good Nose

BY ROGER PINCKNEY

It was two days after ice-up and we were swabbing down the guns. Wind rattled the windowpanes, the woodstove popped and crackled, and Hoppe's No. 9 hung heavy in the air. Chris ran a patch down the bore of his battered double while I poured each of us a healthy dram. The tincture of wood smoke in the icicles I broke from the eaves made good whiskey even better. Porgy was coiled by the stove, but he slept with one eye open, the way he always did when anybody did anything with guns.

This was all up in Minnesota, a good long while ago. Porgy was a Newfoundland, 140 pounds. Broken to harness, he'd pull the boat to the water, fetch up the birds, retrieve the decoys, and drag the boat back uphill when we were done. If only I could have taught him to wind up the strings.

"Hey, Porgy," I hollered.

In an instant he was by my side. There was a sack of decoys in the corner. The green head of a mallard protruded from the top.

"Decoy, Porgy?"

His eyes told me he knew the command.

"Go fetch me up a mallard." He worried it from the sack, came back with it, his tail wagging, the anchor rattling along the floor.

Next decoy up was a bluebill. "Now fetch me a bluebill."

He did. Ditto for a wood duck.

Another mallard on deck. I asked for a teal. He brought the mallard.

"What the hell, Porgy? Didn't I teach you better than that?"

Halfway to my knee, he recoiled as if I had slapped him. He dropped the decoy, cringed, and rolled on his back, four doughnut-size paws in the air.

Chris raised one eyebrow, swirled the liquor in his glass. "I believe you and Porgy are playing a card trick on me."

Card trick indeed.

It was bacon, beans, and kerosene in those days. Porgy kept rabbits and coons away from the garden, found downed deer when I could not. Ten below zero, he'd sleep beneath the meat pole, exploding in baritone outrage whenever coyotes or wolves dared venture into the yard. When winter clamped down and the snow piled to the windowsills, Porgy pulled the groceries home on an L.L. Bean folding sled. When I lugged armloads of firewood onto the porch, Porgy would dog my bootsteps, never happy unless he could carry wood, too, if only one stick at a time.

Things up there did not work out as I'd planned. The woman got the farm but I got Porgy and you'll never hear me complain. I sold what I could, gave away what I could not, threw what was left into the truck, and lit out back to where I had come from. Porgy took up the whole second seat.

I was heading south with the finest company. Leif Eriksson carried a great black bear of a dog on his longship when he explored Newfoundland in the eleventh century. In 1804, Lewis and Clark took a Newfie on their Voyage of Discovery, astounding various tribes all the way from St. Louis to the Pacific and back. Another Newfoundland rescued Napoleon Bonaparte when he fell overboard escaping exile on Elba Island. And the infamous Lord Byron claimed he bedded 250 women in a year during various Italian excursions but wrote one of his best-known poems to Boatswain, his Newfoundland dog, dead of rabies at age five. Byron nursed him to the end, rabies be damned, and never got bitten.

Porgy took well to salt water, though he could never quite understand why it did not taste like the lakes he had known as a pup. He developed an affinity for dolphins and would wade in among them whenever they lolled in the shallows. He'd get nose to nose with the river otters that lived under the floating dock and never made a snatch at one–good thing, as an otter has all the social graces of a chain saw with a stuck throttle. Summers were a little tough. I kept him shaved till he looked like some oversize goofy Lab, and I spent more money on his hair than I ever did on mine.

But it was a string of female companions that Porgy wouldn't tolerate. He had no use for the tattooed wonder from New Jersey I met online. I forget her name. He didn't like Mary Ann, either. She was one of those fire-walking women–you know the type, those perpetually wounded souls who self-actualize by walking barefoot through hot coals at midnight. And down at the beer joint when I'd get a bit soused and try to dance with the local gals,

Porgy would circle and bark. And if I wouldn't stop, he had the good sense to nip me, instead of one of them. I'd just about given up on women when Susan came for a weekend. And when she left and Porgy carried one of her shoes around for the next three days, I knew I had finally gotten it right.

Of all his gifts, this was his greatest. It took us a year or more, but Susan and I eventually threw in full-time. She came with a quick bright smile, neon blue eyes, a giving heart, and an extended family numbering in the dozens. Porgy passed judgment on each of them, found all of them good. And they surrounded me with a love like I had never known. Except from Porgy.

He disappeared right before Christmas three years ago. I walked, I called, I cried. Two days later, there came a faint yipping from beneath the house. Struck by a copperhead snake, he had come home to die. I buried him where the digging was easy.

Oh, Porgy! You never let me test Lord Byron's score, and I never wrote you a poem.

But you have given me one good woman. And I have written you this.

Scents and Sensibilities

BY JULIA REED

A few days into January 2013, I read a blurb in the *New York Times* about Dennis Lehane's beagle Tessa, who had escaped from the writer's Brookline, Massachusetts, house on Christmas Eve. With a determination (and desperation) that I understood entirely, Lehane tried everything to get her back. He took to his Facebook page to enlist help from his fans and accepted the services of a San Francisco psychic. He offered to name a character in his next novel after "anyone who gets her back to us," set up a special phone number so people could call in with sightings, and promised to ask no questions of anyone who might bring her back. For months, I followed the story obsessively, hoping each morning for news of a happy ending. Alas, more than a year later, as I type, Tessa remains in the wind.

Among the many reasons that Lehane's loss almost killed me is that in Tessa's photo she looks exactly like a female version of my own beagle, Henry, complete with seductive dark brown eyes and freckles on her legs and belly. And then there was Lehane's description of Tessa as "smart, fast, and immeasurably sweet," which is Henry on the nose. Mostly though, there was the overwhelming sense of "There but for the grace of God (and, in my case, a holiday turkey and a bag of hamburger buns) go I." Or, more to the point, goeth Henry.

For a dog who often exhibits profound separation anxiety, he takes astonishingly swift advantage of every opportunity to escape. Just a few days ago, for example, I left him—for two hours, maybe—in a very nice room at the dog-friendly Loews Vanderbilt Hotel in Nashville. When I got back, I found him seated in the center of the bed, his head thrown back, emitting a soft and steady bay that was heartbreaking in its sorrowfulness. Then, of course, as soon as I opened the door again, he was down the hall like a shot. (We were on the "hospitality floor" and there were complimentary snacks within sniffing distance.)

Henry does not so much make a willful break for it as automatically follow his nose, which is his one supreme ruler. When humans, who have five million scent receptors concentrated in the backs of their noses, walk into a kitchen with a pot of stew on the stove, they might say to themselves, "Hmm, nice, smells like beef stew." A beagle, who has a whopping 225 million receptors in the back of his or her nose (the membranes of which are roughly sixty times the size of those of a human), will say, "Oh man, cool, smells like carrots, chuck roast, potatoes, some onion, maybe a sprig or two of parsley." This is why you should never try to smuggle a piece of exotic fruit from South America into this country, or worse, some Serrano ham from Spain as I once did: The Department of Agriculture trains beagles to suss out fifty distinct contraband smells.

With a nose that attuned, there's no choice but to follow whatever it identifies. There's also the fact that beagles have been bred for many hundreds of years not just to sniff something out, but to stay on it for as long as it takes. Alexander Pope's translation of *The Iliad* features a line about a beagle and a deer shot by a "distant" hunter: "Thus on a roe the well-breathed beagle flies, And rends his side, fresh-bleeding with the dart." But it hardly takes a roe—or even a rabbit, which is what today's beagles are generally bred to go after. A squirrel, an ancient chicken bone, pretty much anything will do the trick. Most electric fence people refuse to sell their product to beagle owners—a shock is nothing to a beagle on a hunt. Henry once broke free of my fairly firm grip in order to consume a roast beef po'boy (along with the foil in which it was wrapped and the paper bag that contained it) lurking in my neighbor's ground cover.

The single-minded quest for bunnies, beef stew, or indeed roast beef on a roll is what makes Henry so easy to lose, but also (so far, knock on wood) to catch. On his very first Christmas, we didn't even know he'd gotten out until my next-door neighbor—whose wife, apparently, gets the holiday feast on the table a tad earlier than I manage to—called to report he was in their dining room.

Then there was an especially scary night in Seaside, Florida, where my mother has a house. I had a dinner party for my friend the painter Bill Dunlap after an opening of his work. At one point, before we all sat down, Henry took an opening of his own and nosed the back door open. By the time we realized it, the sun had gone firmly down and I was firmly hysterical. There are few streetlights in Seaside; the town is bounded by a busy road and lots of dark wooded acres. While everyone fanned out with flashlights, Dunlap kept telling me to calm down, that the dog would find his way back. And I kept telling him, "You idiot, he's an effing beagle, they don't come back," except in language far, far more pungent. The more Dunlap reassured me, the more I yelled, which turned out to be a good thing. Two doors down a woman stepped off her porch: "Did you say beagle? We have a beagle in our kitchen."

As it happens, the woman's husband had been grilling burgers for the family, so Henry had simply invited himself over. By the time I arrived, he'd eaten an entire bag of buns and was on the floor good-naturedly allowing the two-year-old son to roll and unroll his silky ears repeatedly. Such is Henry's considerable charisma that the nice people did not complain a bit about being forced to eat bunless burgers, and the boy wailed like a banshee when I removed his new playmate from the premises.

That's the thing about Henry. He brings out the best of even the worst people. Dog books regularly describe the breed as "cheerful," but Henry goes one better. When he's out and about, he can be downright merry, and no matter what he's doing he's amusing, which is the trait I also most value in people. When we walk, passersby roll down their windows to holler out some version of "Great dog!" and kids follow us on bikes. A tattooed, shirtless workman with a cigarette in one corner of his mouth once dropped to his knees to coo over him, and last summer, in Seaside again, we encountered an entitled adolescent trash-talking a girl to his companions. But as soon as he saw us, he became the soul of politesse: "Ma'am, that's a beautiful dog."

When our house was broken into several years ago, the thief

took the time to lock Henry up in the laundry room rather than simply clubbing him on the head or worse. The man, who turned out to be a twenty-year-old crackhead, had come in through the kitchen window, and I would bet money that Henry leaped up on the low counter to welcome him with his usual enthusiasm—my neighbor thereafter referred to him as "the burglar's assistant." Because that's another thing: Henry has never met a stranger. I used to feel really, really gratified by the greetings he gives me whenever I've been away. He runs around in crazy tight circles, presses his head submissively into my thigh for long moments, and then jumps on the hall bench to lick my face before jumping off to run around in circles again. Now that I have seen him go through pretty much the same routine (without the thigh thing) for everyone from the UPS driver to an Orkin Man he's never seen before, it has lost a tiny bit of its poignancy but none of its charm.

He might hurl himself into the arms of a bad guy, but the one thing Henry will not do, ever, is come when he is called. Occasionally he will come to the word "treat," and invariably he will come to an actual treat. It's not that he doesn't know his name—every time he hears it, his ears perk up ever so slightly. But when you call him by it, he either sits or stands stock-still and simply stares. When this happens, it always looks like there's a lot of stuff going on in his mind, like he might be turning over a series of options. But I have no idea what they are. Henry may well be the most outgoing dog who ever lived, but he has always kept his own counsel.

As much as it sometimes irritates me, a lot, I have a grudging respect for this air of mystery—which is, after all, in way too short supply among most two-legged folk. And for all his secrets, there are a few things I know for sure about him. He will eat anything.

The crazy Cajun from whom we bought him (at a tiny seven weeks for a paltry hundred dollars) told us he'd been feeding him powdered eggs, so there was no place, I suppose, to go but up. As a puppy, he ate the better part of a pair of Carolina Herrera pants, a pair of Manolo Blahnik shoes, and the binders of all my cookbooks on the bottom shelf. He once ate George Washington's face out of the center of a dollar bill, and when the dog sitter replaced it with a photo of Henry himself, it looked surprisingly convincing—he adopts a slightly regal air whenever he poses for pictures, which is often.

He knows to the second when it is five o'clock, the time of his afternoon feeding. He prefers down pillows and will snuggle up to anyone wearing fleece or cashmere. The squeaker in a stuffed toy has no chance against him, but he is scared to death of cats. Between the two, he will choose the company of toddlers over other dogs, whom he doesn't really understand. He has a vivid imagination (the only time he ever barks is in his dreams).

Finally, while I am pretty sure he loves me, the thing I know the most is that I could not love him more if I tried. He captured my heart completely in the yard of a barefoot Cajun septuagenarian who "had his papers somewhere around" but couldn't quite put his hands on them. Henry doesn't need papers to prove his nobility to me. But he does have a chip and a tag with my phone number. So if you find him on his own, please call me. I can't name a character after you because I'm not (on purpose) a writer of fiction. But I will give you whatever else you ask me for.

Ole No. 7

BY ACE ATKINS

He wandered up to our farm in Paris, Mississippi, late one night, skin and bones, starving to death, a hunting dog who'd lost the trail of the hunt, as hounds sometimes do. At the bottom of Faulkner's county, Lafayette, people run deer with short-legged beagles, mutts, and sometimes fine hounds. This dog was a fine hound, an American foxhound, tricolor and long legged. He didn't trust me at first. It took a bowl of food every night for nearly two weeks before he'd let me get near him.

I would watch him from the shadow of our porch, waiting until he'd heard the rattle of dog food to skulk to the bowl, devouring the food in huge gulps and returning to the woods.

You didn't have to be an animal expert to know he'd been mistreated, most likely left in a filthy pen until used in deer hunting season. He looked old, my wife, Angela, and I thought, with worn teeth and a bad coat. Turns out he wasn't old at all, but as Indiana

Jones once said, "It's not the years, honey, it's the mileage." I imagined he never had a name, probably just a number, and so I called him Ole No. 7–a good name for people who liked whiskey and for a dog who'd wandered into a lucky situation.

Part of us kept hoping he'd fatten up, rest up, and then find the trail of his buddies and move on. Over the eleven years we lived in Paris, we rescued a lot of dogs. We did not actively seek them out, but most of the time they found us, either by managing to stand on some lonesome highway in front of our car, or by wandering right up to the house as 7 did. We lost count at more than a hundred that we fixed, vaccinated, dewormed, and found new homes for all over the country. The first came eleven years ago: five puppies born under a trailer down the road to an endearingly ugly boxer named Toyota. The last–we hope–came just months ago, even though we weren't living there full-time: a litter of six cuddly, barely weaned black Labs left to fend for themselves in thirty acres of logged-out wilderness beside our farm.

So many others came in between. There was Lena Grove, the shy, wandering mutt who'd walked a "fur piece" to us, the cattle mixes Hud and Alma, bird dog mixes Gumbo and Beaudreaux. Gumbo became the companion of the journalist Rick Koster in New London, Connecticut, and spent the rest of his life taking long walks on the Atlantic coast, leaving this earth with a first-class obituary in *The Day*. So many others that sometimes we'll pass a spot on the highway and say, "Remember June?" "Remember Murphy?" "Remember Hank?" "Remember Callie?" and those weeks and months or even a year that they passed in our care are blurred by that one moment that they trusted us enough to approach, eyes pleading, starving and in trouble. A friend once commented that we had been marked by dogs the way Depression-era hoboes marked the houses of people who couldn't turn away a hungry man.

The Atkins farm, which we called Carrefour (crossroads), at different times burst with up to sixteen dogs. Even though we've moved into nearby Oxford, they still keep coming. A skinny greyhound mix that Angela calls Loretta Lynn was the latest hard-luck case: Starving, pregnant, covered in fleas, she appeared one day, dug herself under the fence, and never left.

Ole No. 7 was different than most. A purebred, he was no scraggly mutt. But it took a huge vet bill that included treatment for raging heartworms, among other maladies, before he began to show it. The white in his coat became whiter, the dull look in his eyes brightened, and most of all, he found his place at the farm, believing his job was to protect it from wandering coyotes and anyone who might get too close to the house, a job he carried out in a new regal stance–head lifted, tail high. The stance became his hallmark.

We never put 7 in the canine underground railroad, joking that no matter where he ended up, he'd most likely walk back to us in Paris, Mississippi. Never confined (except during his heartworm recovery), he roamed the pastures and woods, took long swims across our bass pond, and joined the family for winter bonfires. During a walk, you didn't have to look for him. You could close your eyes and reach your right hand down by your side and find his big head and wet nose as he trotted close. He became guardian and pal to our two children, letting them try to ride him like a pony and pull his velvety oversize ears.

Not that 7 was civilized and easy. He still chased deer, and sometimes with the help of his ever-present companion, the rescued bird dog Patches, he caught them and dragged them back to the farmhouse. He could also be a real asshole at feeding time, refusing to let any of the other hounds near the trough until he'd gotten his fill. But after all he'd been through, how could you blame him? He became as much part of the farm as the bonfires and the front porch and the bass pond and the old barns. When I would drive back home on our gravel road, 7 and Patches were there to greet me at the mailbox, following my truck all the way up the drive, baying and howling their greeting.

When he got sick, as all dog stories must end, we knew first by his absence at dinnertime. Angela found him hobbling, his back legs swelling to four to five times their size. We, along with his vet, thought he'd gotten into a nest of water moccasins and that it would only take some good care to get the poison out of his system. But the swelling never left. We learned the poison wasn't from snakes, but from his own body–a fast-growing cancer had nearly blocked his bowels. Ole No. 7 was dying.

I saw him one last time in the vet's office before decisions had to be made. He again lifted his big old head as a welcome, and I rubbed his ears and called him my old friend as his eyes said the same to me. I was far away from Mississippi when he had to be put to sleep. I had just stepped off a train at Penn Station and had to go through a series of meetings not being able to talk with anyone who would understand the loss of such a noble, fine dog.

We still have the farm but don't live there anymore. It still seems odd and quiet out there without his baying or seeing his tall, upturned white tail to lead my truck back home.

That fall, we scattered his ashes at the farm in the back acres he protected from coyotes, up at the bass pond where he loved to swim, and down the long gravel road where he loved to trot and sniff and explore.

He was one of us.

A Dog to Lean On

BY MARTHA J. MILLER

My older sister, Ashley, and I begged our parents for a dog for the better part of two years before they finally gave in. I say "they" but in truth, there was only one "no" between the two of them and it belonged to my stepdad. In his mind, dogs smelled, drooled, chewed furniture, peed on the living room rug, and puked under the dining room table.

"Y'all will have fun with a dog for about a week," he said, "and then I'll end up having to take care of it." "*Nu-uh!*" we chimed at him together. "A dog is a great way for me to learn responsibility!" I added.

I could bullshit with the best of them, even at twelve years old.

Mom wanted a dog just as badly as we did ever since our first family dog, a standard-size dachshund we'd named Tildy, had died a few years back. And by die I mean that Mom accidentally hit her with our 1988 Oldsmobile station wagon on a return trip from the grocery store. What's worse is that Ash and I witnessed the horror in real time from the front porch steps of a neighbor's house.

This all went down not long after my dad moved out and just before she, a newly single mother, moved Ash and me out of a hill town in West Virginia and into a small rural town outside of Richmond, Virginia, for a new start. A second beginning that would eventually lead to a second marriage and a stepdad I liked. Except for his distaste for dogs.

As three refugees of divorce, Mom, Ash, and I were coming fresh off a rough couple of years, and we figured we were about due for a good one. A new dog to love, we thought, just might make it so. One night, I pulled out all the stops at a good old-fashioned family meeting. But in the case of *Canine Enthusiasts v. Fatherly Skeptic*, the judge did not rule in my favor.

Heartbroken, I retreated, but not without first issuing a final pathetic plea. "Can I at least get a hamster?" The next day after school, the phone rang and I answered. It was my stepdad calling from work. "You're only twelve once," he said. "If you want a dog, you can have a dog." I never talked of hamsters again.

We found Harvey utterly by accident. Mom spotted him first, sitting in a wire cage, just steps outside the door of the local Ukrop's grocery store. He was on display as part of an adoption event being held by a community rescue organization, and we were drawn immediately to his comical good looks. Half basset hound, half Labrador retriever, or so they guessed, he had a coat that was coal black save for a bright white starburst of fur on his chest and splotches on his feet, as if he'd pranced lightly through a puddle of white paint. Looking like a Lab that was cursed with stubby basset legs, Harvey rode low with a tail that thumped the floor loudly on approach. "He's never known a stranger," the rescue volunteer said. "I'm fostering him in my home right now, and he just sleeps on my feet all day while I sew." And with that, we loaded our groceries and a squat little hound dog into the back of our car.

Once home, Harvey set about the business of making himself comfortable. He searched out the best nap locales, under my parents' bed for the silence and under the dining room table for the daily menu specials. He introduced himself to the neighbors, often finding his way through an open patio door and onto the dance floor of the local country club or making the rounds for his daily treat collection.

And though he ultimately proved to be of the "lazy canine" variety, every so often he'd catch a whiff of something and the hound instincts would take over. Running as fast as his nubby legs could take him, he'd disappear into the woods to wallow in the creek beds or roll in the warm cow patties, which speckled the pasture of a nearby farm. Just as the sun began to wane and the cicadas started up, he'd trot home, head bowed and looking like he'd been smeared from snout to tail in peanut butter. A predicament that only a garden-hose bath in the driveway could dare to rectify.

In those early years, Harvey was largely a dog belonging to the female side of the house. My stepdad didn't seem to mind him, but there was a certain indifference to his presence, aside from the handful of times we'd catch him secretly warming Harvey's canned dog food in the microwave. Then my sister and I went away to college, and my highly driven, exceptionally successful stepdad found himself with a family construction equipment business that he no longer owned and plenty of idle time at home.

My stepdad had always been the consummate provider, a classic strong and silent Southern-man type for whom identity and profession were one and the same. When he was a young boy, he accompanied his father on drives up through Virginia's Shenandoah Valley to see the blasting and building of I-81. And it was on those drives that he first came to know his future. Taking over the reins of his father's business was his legacy, his birthright. But birthrights don't always stay with their rightful owner.

In a contemporary economy where the competition was no longer the guy next door, but big corporations with slim overheads and long arms, he and his father opted to do what they could for their employees and sell the rest. So at fifty-one years old, my stepdad was at home with the bantering of daytime TV, the chore of picking another color for his parachute, and the company of a smelly, overweight hound dog that he never really wanted.

Unemployment and forced personal rebirth can be emotionally jarring for anyone, and my stepdad was no exception. However, as a man of few words, he never spoke of it. Not to any of us anyway. Instead, he spent his days closed up in his home office. And then somewhere along the way, Harvey decided to join him, snoozing his doggy days away in a dim corner near the bookshelves. The two of them became an inseparable duo seemingly overnight.

On a return visit from college one weekend, I witnessed the transformation myself. I watched as my stepdad exited his office to grab a beer from the fridge. His footsteps across the house were followed soon after by the *click clack click clack* of Harvey's toenails on the hardwood floor. A beer for the man, a treat for the dog. Back to the office they went. Man and dog coexisted largely in compatible silence, broken only by an occasional remark or question tossed into the air by the man. "Hey, buddy, wanna go outside?" The sound of Harvey's tail on the carpet, *thump thump thump*, signified an affirmative reply.

Simple communications regarding walks, bathroom breaks, and treats were all I had the privilege of witnessing. But the improved, even cheerful, demeanor of the man and the doting affection of the dog hinted at potentially unseen exchanges that went deeper. Words uttered by the man that he could never find a way to say to his wife and stepdaughters, but that crept out in the lonelier hours of the working day. Days of anger and healing witnessed only by the soulful, loving eyes of the dog.

Many years have gone by since then and Harvey is no longer with us. Old age eventually began to claim his eyesight, and after a 3:00 a.m. tumble down the stairs, which required my stepdad to carry him outside for a bathroom break, we knew it was time. My mom and Ash handled Harvey's teary yet peaceful end on their own. Uncomfortable with the raw and ugly emotions that life can bring, my stepdad did not make that final trip to the vet.

Some may not understand why he couldn't be there, why loading Harvey gently into the back of the car and saying his goodbyes right there in the driveway was the very best he could do. But I'd hazard a guess that Harvey knew why, and the hound's knowing is the only one worth counting that day.

For years after Harvey's death, my stepdad claimed he would never own another dog despite Mom's desire for one. He returned to old mantras, saying, "Dogs are too much work." A tough-guy remark we all knew to speak more of fear and personal grief than actual inconvenience. Until one day a few years back when he reluctantly accompanied Mom to a dog rescue adoption event. They returned with another squat hound mix, which my stepdad named Sidney, after his brother. Sidney is more beagle than basset, but with just as much hound character as Harvey.

Man and dog are once again inseparable. Sidney sits like a miniature business partner in the front seat of my stepdad's car, accompanying him on real estate showings and errands. At home, the dog snoozes while the man works at his new real estate career or reads the latest online predictions about Virginia Tech's upcoming football season.

"I agreed to get another dog for your mom," he likes to say. "I know how happy they make her."

Like me, he can bullshit with the best of them too.

Dog Gone

BY ROY BLOUNT JR.

You know how to test the proposition that a dog is man's best friend? Lock your dog and your wife in the trunk of your car, drive out into the woods somewhere, and let them out. Which one do you think is going to be glad to see you?

That's just a joke, of course. My wife's favorite, I might mention hastily. But every time I tell it, I feel a pang.

I have written two dog books, in dogs' poignant voices: *If Only You Knew How Much I Smell You* is one of them. I've written a country song title, "You Know How a Dog's Leg (Twitches When You Scratch Him on the Stomach; Well, That Is How My Heart Is Doing, Darling, Over You)." The only country song title I know of that has a semicolon in it. I think less of someone who doesn't have a dog.

And these days, that someone is me.

There's old dogs and young dogs
And big dogs and small,
And then there's what I've got:
That's no dog at all.

My dog is nameless,
Won't come when I call,
Won't miss me or kiss me
Or run for a ball.
But none of that's his fault–
He's no dog at all.

Yes, I *want* a dog–how superficial and un-down-to-earth do you think I am? I want *your* dog, probably. Don't ship him or her to me, though, because that would make you even more fly-by-night than I am.

We always had dogs when I was growing up, and when my kids were growing up. My wife had dogs as a child and in her previous marriage. But for the twelve years we've been together, she and I have been–let's not mince words–dogless.

Oh, sure, in our yard we have a concrete dog and a terra-cotta dog, and on the kitchen table we have a stuffed dog who, when you press his foot, sings "I feel good!" in the voice of James Brown. But that's it.

We travel so much. Even if we knew someone who would keep our dog when we are popping into the city or hitting the road, our dog would be *pining, howling, worrying.* How can you enjoy a trip if you know it is breaking your dog's heart?

I know. I know. What kind of cold, sorry, un-regular people would rather "travel so much" than have a dog? So stay home, you're saying, with your dog.

I can't stay home. People won't buy a book unless they hear you on the radio frequently and then you come to their town and entertain them for an hour (along with the much greater portion of the audience that doesn't happen to be in the market for a book right now) and then sign the books with special shout-outs to the several people they're planning on lending them to. Also—what, you don't want me to spend time in New Orleans? And I'm writing this in Savannah, where we just finished eating eleven different dishes and biscuits and cornbread family-style at Mrs. Wilkes Dining Room. I guess you think we could eat like that in Massachusetts?

Okay, you say, then carry your dog or dogs with you, as Amy Tan, your fellow member of the authors' rock-and-roll band the Rock Bottom Remainders, has done, in her purse, all over the world.

No, my dogs have always been too floppy and vocal to fit serenely into a purse. If you had tried traveling with any of my dogs

in a satchel, even, you would look over and see that the satchel was rolling in something nasty.

I'm about to mist up.

As it happens, my wife is smitten with a Havanese dog that comes to her exercise class. Having visited Havana, I would expect a Havanese to be made up of parts from eight or ten rusty old dogs, which sounds pretty good to me, in that I have always had hybrid dogs, from the pound. No, Joan says, this dog is purebred. But he is the *funniest* dog, she says. She says there's a place in South Carolina–not a puppy mill–where we could go look at Havaneses. A dog who comes from South Carolina and likes a good laugh can't be too lah-di-dah. And if we do go look, you know what's going to happen.

Running Mates

BY VANESSA GREGORY

The dog walks the road leashed to restrain his exuberance. We are vacationing in northern New Mexico for a few weeks, trading the fecundity of Mississippi for the expansiveness of high desert. He trots ahead of me, stopping to inspect a culvert, lowering his head and bringing his shoulder blades together on his back. I pull lightly at the leash and he plunges forward again, leading with the nose.

Frank is a handsome animal with a coat like polished walnut and the lean muscles of a sporting dog. At home in Oxford, he elicits compliments from men on the street. They think he's a gentleman's dog. In reality, we found him at a county animal shelter, where he shared a metal pen with a hyperactive puppy. We had lost his predecessor, a gentle giant of a Labrador retriever, to cancer barely two months earlier. I wasn't ready for a new dog, but I couldn't stand living without one either.

Before he'd landed in dog jail, Frank had been a fixture at the local high school, using his expressive brown eyes to cajole snacks from the kids. When the school closed for winter break, administrators called animal control. The shelter Frank ended up in had so many good dogs I could hardly choose. What about the yellow female smart enough to unlatch her pen? Or the fuzzy black mutt who scarcely made a sound? Finally, the kennel attendant settled it for us. He said that Frank was his favorite, and that he'd been there a long time. So we paid thirty-five dollars and took the dog home. Frank's body shows evidence of Labrador retriever, but he won't swim and is a halfhearted fetcher. We'd soon learn that his soul is something else.

This vacation he has a routine. After the road, we return through the gate. The house we're renting sits alongside two others on three fenced acres. We stand at the end of the long gravel driveway. Frank sees things I can't see. Or smells them. Or hears them. Or all three. His ears dip forward and he focuses on some point in the middle distance, past the grasses and anemic apple trees growing in the yard to our right.

I tell him to sit and he obeys, slowly, his rear drifting downward in increments, his attention still fixed elsewhere. But he knows what's coming. The muscles in his shoulders twitch. I reach down to release his collar and in the same instant he bursts forward like a sprinter out of the blocks, bounding past the sagebrush in long, graceful strides. The earth gives off a quiet thud each time his paws hit the dirt.

We don't know where Frank came from or the exact composition of his pedigree. But we know he is a runner. An athlete. A being who becomes his fullest self in these moments spent striding across hills and fields. We discovered this a few weeks after bringing him home, when we decided to unleash him on the wooded trails south of the University of Mississippi campus. Frank trotted at our heels for twenty yards before darting away. Soon, we could neither see him nor hear the clink of his tags. We shouted but it didn't work. We thought we'd lost our new dog. Twenty minutes later, Frank was back, panting and wagging with a look of satisfaction in his eyes.

Since then, my husband has clocked him racing alongside an ATV at thirty-six miles per hour. He leaves other dogs huffing in his wake and shuns the easy trail in favor of the most intriguing route. When we hike in the forest, he is behind us, in front of us, to the side of us. He is clearing a log in a confident leap or charging up the side of a ravine or loping out of sight. He has twice run so hard that the pads on his paws have peeled off. We've started buying a protective product for sled dogs, a sort of paw wax, to smear on his feet.

In our early days together, I wasted countless afternoons trying to teach Frank to come on command. But he isn't the kind of dog that will trade his freedom for a liver snap. Eventually, I learned to trust that he'd return on his own.

Here in the open terrain Frank can achieve full gallop. He sails in an arc toward the east, where the green Sangre de Cristo Mountains rise from the brown plains. Let's be clear: There are rabbits in this yard, and Frank will chase them. He will roust them from the bushes and track them in frantic circles, nose to ground and tail to sky. He is equally vigilant against birds, happy to charge at the finches pecking seeds beneath the feeders. But that is not what is happening now. This is about speed and power and motion. He has no destination. He is just running to run.

In twenty minutes, Frank will be sated. He will trot inside, tail wagging, to consume a bowl of food and drain his water dish. Then he will lie on his side on the cool concrete floor, eyes closed, the barrel of his torso rising and falling as he sleeps. He will be as still then as he is fluid now.

For the moment, I watch him. It will take about twenty seconds for Frank to disappear behind a cluster of sage-colored bushes or to glide around the corner of the caretaker's house. But while I can still see him, I am transfixed by his strength and the joy he finds in his body. Does the gravity-bound canine dream of flight? This is close. Out here it's obvious his ancestor is the wolf. He's as much a part of the landscape as the peaks and dust and sky.

Frank is still a young dog, maybe four years old at most. I don't like to imagine a time when his joints might stiffen and his sprint slow to a saunter. Some dogs like to run, and other dogs need to. I think that somewhere in the depths of his beast's brain, he'll remember his former swiftness and he'll miss it. Today, though, he is a marvel, every sense alert, making full use of the body he was given.

I am also a runner, although if Frank could speak he'd probably call my plodding gait a jog. Sometimes we run trails together and I look up to see him paused with a front paw lifted. He stares back at me expectantly, as if my turtle's pace were a conscious ploy to slow him down.

Runners cannot explain themselves to nonrunners. Critics say that our knees will grow brittle and fail. They question the utility of sweat and exhaustion and ragged breath. Maybe there's no good explanation for why we do it. Maybe the instinct comes from somewhere without words. But I can tell the naysayers this: Some days I follow Frank's lead. I abandon my measured trot and accelerate, stretch my legs until the muscles ache, hear my feet strike the earth, swallow air until my throat burns. I become an animal sure of its place in the world.

Literary Canines

BY CURTIS WILKIE

Among the many things that journalists can't keep, it's been said, are pets and plants, and for most of my career I was proof of that axiom. We travel too much. But after I moved to New Orleans in the 1990s and began to settle down, a dog sounded appealing.

Although I never hunted, I thought yellow Labs, with their soulful eyes and long, floppy ears, were especially neat and asked my friend David Crews, who was looking for a Lab for himself, if he could find one for me, too. He located a kennel in South Mississippi and brought away a pair of six-week-old puppies. David took the female and named her Lion, after the dog in the Faulkner story "The Bear." I got Lion's brother, all ears and paws and energy. Not to be outdone, I seized on a literary name, too: Binx.

I wanted to name my pet for a fictional character with a New Orleans background but dismissed Ignatius J. Reilly and Stanley Kowalski. Neither name sounded lyrical. But Binx had a nice ring to it, and there had been times in my life when I identified with Walker Percy's creation, Binx Bolling, and the fugue state he moved through in New Orleans in *The Moviegoer*.

So Binx came to live on Burgundy Street in the French Quarter, and I discovered that in lieu of game birds he was quite willing to snack on shoes or ties or handles of cooking implements. Or, as a special treat, the Oriental rugs I brought home after years in the Middle East. Yet it was impossible to get too angry. When I yelled, Binx would lower his tail and look at me sadly with his big, deep brown eyes. And I would be the one who felt guilty.

We fell into a comfortable routine. Cabrini Park, across the street from my house, was known in the neighborhood as the "dog park," and in the late afternoon Binx and I joined dozens of residents, bearing their cocktails and their own pets. It was a daily festival of greyhounds and shepherds, dachshunds and poodles, AKC purebreds and rescue animals.

Almost every day we'd also walk along the levee. If it was especially hot, I'd let Binx dip in the river while I held his leash. Once, as we strolled in front of Jackson Square, an overheated Binx jumped spontaneously into a water trough for the mules that tow tourist carriages.

"Get that goddamn dog outta there," one of the carriage drivers shouted.

"Buzz off," I barked back, or words to that effect. "My dog is a helluva lot cleaner than your mule."

Better swimming was available across town in Bayou St. John near City Park. I'd fling a rubber chicken into the water, and with a running start Binx would dive in with a big splash to retrieve it.

There were real ducks to chase, too. But I watched out for alligators; I had nightmares of Binx in the jaws of a predator.

At home, we'd play "hockey." With a tennis ball, I'd kick slap shots and Binx, the goalie in the kitchen doorway, would swat them down.

Binx became a member in good standing of the Mystic Krewe of Barkus, a rump carnival group designed to mock the swells from Uptown and their elite Mardi Gras floats, and pseudo-royalty. Barkus staged an annual parade with hundreds of dogs marching through the French Quarter, led by a king and a queen selected from the pound.

Binx acted as his own chick magnet. One Barkus spectator asked if he could bring his yellow Lab, Annie, to Burgundy Street for a date. Liking the idea of another puppy that might materialize from this union, I invited Annie over. But as soon as Annie found herself in the courtyard with Binx, she submerged into our fountain, leaving only her head above water. Though Binx stood by, panting, she held that pose all afternoon.

"There are lots of other girls out there," I assured Binx.

Before long, my neighbor Phillipe LaMancusa proposed a mating with his sweet little Lab, Ginger. We often encountered Ginger in the neighborhood. Because she was blonde and originally from California, I called her Surfer Girl. The pair proved compatible, and on Mardi Gras Day 2000, in the French Quarter, Ginger delivered a litter of eight.

Empowered with "the pick of the litter," I inspected the squirming mess of Ginger and Binx offspring and chose the rowdiest, branding his belly with a black marker. When I returned to fetch my puppy the next day, he had licked away the ink. But I recognized him by a single pink toe I had noticed the day before, and I knew I was getting a rascal.

Following "literary tradition," I named the puppy Willie, after the writer Willie Morris. Before his sudden death a year earlier, Willie often dropped by my house when in town. A noted dog lover, Willie was fond of Binx, so the name seemed proper.

Willie inherited Binx's appetite. While I was away that summer, covering the 2000 presidential campaign, Willie consumed an expensive leather jacket, more shoes, and a leather-bound book with photographs of my daughter's wedding. He ate plugs out of two more Oriental rugs.

While traveling, I checked in by phone with the house sitter. She wailed, "He's even after the leather couch." She asked for permission to keep Willie penned outside, but I told her to leave him in air-conditioning. New Orleans summers were too great a penalty to pay for the havoc. The damaged goods were material possessions, I rationalized, and Willie far more valuable. When I finally arrived home, I discovered he had chewed off most of the leg of a cabinet.

The next year, he gnawed away the cover of David Halberstam's book on the Balkan conflict. The author later inscribed the remains, "To Willie Morris Wilkie, with admiration for the quality of his taste."

Eventually, the three of us moved to Oxford, Mississippi, and "my boys" seemed just as happy there. They had a gentle, clear creek in which to romp on my son-in-law's land in nearby Taylor. In town, they liked to strut around the Square. We were early risers and usually joined predawn joggers on the streets. If I was not awake by 5:00, Binx and Willie got me up, butting the mattress with their big square heads. I think they could read digital clocks, because they reliably stirred at 4:44. They actually made the act of rising a joyful exercise; they looked like they were laughing.

Though no longer puppies, they were still mischievous. When my former roommate, Joe Fenley, flew in from California for an Ole Miss football game, he stayed at our place. It was the seven-overtime game with Arkansas that lasted until midnight. When we got back to the house, Joe discovered that his pair of Ferragamo shoes had been reduced to soles. He wrote a thank-you note for the hospitality, but noted that "it would have been cheaper to stay at a Four Seasons Hotel."

As the dogs of a writer, my boys wound up in print. Binx was immortalized in *Feet on the Street: Rambles Around New Orleans*, Roy Blount Jr.'s book in which the author described the pleasure of leading Binx during a Barkus parade.

Willie had his own special moment of recognition. I had written a book, dedicated to Willie Morris, and mentioned in a footnote that my dog was named in his honor. Afterward, I got a letter from former president George H. W. Bush, whose campaigns I had covered. "I realize we had huge differences about politics, political philosophy, you name it," he wrote me, expressing delight to learn from the book that the *Boston Globe* and I had been just as hard on Jimmy Carter. (I always enjoyed bantering with Bush; he is a good sport.) The former president added that Willie Morris had been his friend. "He came to the White House. We talked baseball, dogs (Go, Willie!), and life." I preserved the letter before my Willie could devour it.

In 2007, we had to make a major readjustment in our household. I married a wonderful lady from Memphis named Nancy.

She loved dogs but owned, instead, two cats: a regal Persian named Pudding, and Clamp, a feline without any notable pedigree. Clamp was so named because as a tiny kitten she had been rescued from the jaws of a dog. Not exactly a good introduction to coexistence with two big Labs. But a couple of days after our wedding, the newcomers arrived in Oxford in cat carriers.

As I set the carriers in the hallway, I saw Binx and Willie approaching, tails swinging. "Look," I said to Nancy. "They're going to like these cats."

Nancy was not so sure. "I think the dogs' tails are wagging because they've just seen dessert," she said. Suddenly the hallway erupted in a cacophony of snarls and barking, accompanied by hissing from the cat carriers.

For about a week, we kept the dogs and cats in separate rooms. Finally, we decided we couldn't go on living this way and opened the door to the front bedroom, where the cats were stored. Eventually they emerged. Clamp's timorous steps reminded me of Carl Sandburg's fog that "comes on little cat feet." But Pudding fairly swaggered past the dogs, ready to bop them if challenged. Binx and Willie looked on, with amazement. Like most Labs, my boys were actually very gentle. The cats' siege was broken. The dogs and cats not only accepted one another, they developed affection for each other.

On my desk, I have a photo of Nancy handing out treats to all four of them, huddled together beneath her hands.

I'd like to be able to report that we all lived happily ever after. But that autumn, Binx was diagnosed with cancer. On our walks, Binx began to chuff heavily when we headed uphill. I recalled a passage from a Jim Harrison novel where a character pitched "birthday treats" to his dog every day because dogs don't get enough actual birthdays.

Sometimes, when I'm down, I resort to doggerel (no pun intended). On Binx's twelfth birthday that November, I wrote "Walking with Binx."

Flecks of frost glint on the lawn
in the first light of dawn.
Leaves pale and dry as aging skin
scud quickly through the wind.
Stricken flowers sag in the weather;
the bright hibiscus begins to wither.
The dear old dog is 12 today.
He plods, determined, remembering:
Swimming in the creek,
fetching sticks,
racing Frisbees,
chasing geese.
Not knowing this will be his last winter,
he no longer strains against the leash.

When it was time, we wrapped him in an Ole Miss saddle blanket, with his Red Sox leash, Mardi Gras beads, and his Krewe of Barkus credentials, and buried him by his beloved creek in Taylor.

Pudding died, and then Clamp. They both rest now in our garden, under headstones.

In the spring of 2012, we were told that Willie had only a few weeks left. I took him to New Orleans for a "victory lap" around the French Quarter. He remembered the neighborhood. His spirit was indomitable; he refused to die. For half a year Willie lived on, in apparent good health, before he began to fade that September.

Half of his ashes are buried in our courtyard in New Orleans, the rest under a new maple tree in our backyard in Oxford. I used a New Orleans expression for his headstone: "Willie. Every day was lagniappe."

Indeed, every day with my dogs was something extra.

Delta Dog

BY J. M. MARTIN

In the Mississippi Delta there is a proud and indeterminate breed of dog that can be seen from passing cars loping along the highways and levees and turnrows. Their eyes are ghost blue or wild gold or the muddy gloom of the river, sometimes all of the above. Their fur is slick and greasy or long and matted. Their ears, chewed and pierced at the edges, generally stand at attention, expecting some old trouble to get into. Their tails, if there's any tail left, curl up in puckish question marks or drag along, forever tucked between their legs. They generally run in ragtag packs, the most noticeable wildlife around, and when you blow by, they'll stop their loping and follow you with those eyes, like "What are you doing here?"

Life for a Delta dog is free but not easy. When I ventured down there to work at the newspaper in Greenwood, the yard of the old house I moved into came with a dog. No one owned the dog; that was just where it had taken up. It was a shaggy gumbo of a mix, roughly chow and German shepherd, its thick coat matted into long dusty dreadlocks. My landlord, who lived next door, called it Shepherd. My girlfriend, Kate, who taught school in Indianola, took to calling it Pigtail. We wouldn't touch the thing. If it weren't for Tommy, the twitchy recluse who lived in the little house out back and blasted Elvis at night so he didn't have to hear himself think, the dog wouldn't have had anybody. A Delta dog will find its way.

One night in Indianola a friend of ours showed up at a party with a box full of fuzzy blond and brown and mottled pups, an entire litter. Delta dogs are born in ditches and culverts and lean-tos and body shops and, yes, barns. She found these under her neighbor's house, unwanted by the neighbor and abandoned by their mother, a reputed woman of the night named Queenie who left many litters to the same fate. No one immediately jumped at the chance to adopt one. They were cute, but they looked like work. We took them out in the backyard and raced them like turtles. People were carrying them around like personal mascots, drink in one hand, dog in the other. By the end of the night, those drinks working in their favor, all of the pups had become party favors. We were awoken at five the next morning by the yelps of our new responsibility.

You can see anything you want in a Delta dog. Any combination of water dog, coon dog, bird dog, fighting dog, work dog, you name it. The Delta dog is the alpha dog and quite possibly the omega dog, the dog from which all other dogs came and are coming to be.

At first Vernon looked like a golden retriever, distinguished by a white racing stripe running up his snout and little white slippers of paws. Soon other breeds began to emerge: a head shaped like a Lab's, a husky's impermeable coat, the mustache of a wolfhound, the markings of a collie, a spaniel's webbed feet, the bulbous pink nose of a pit bull, a point like a Brittany's. Although various breeds found their way into Vernon, he inherited none of their use. Wouldn't swim, couldn't hunt, refused to even fetch. This seems to be the way with dogs: The less of a damn they're worth, the more to death they're loved.

We did some checking into his lineage. There was Queenie, of course, and then there was his father, a retriever mix named Shop Dog after the garage bay where he resided. To complicate matters, we were told his uncle was a one-eyed husky named Waylon. None of which explained anything.

Vernon spent his Delta days giving in to the same wandering spirit that possessed his cousins. I am ashamed to admit that on more than one occasion we kept him tied up in the backyard. We were young and just out of college and still learning the limits of our humanity. The backyard was inhabited by a sprawling fig tree, which Vernon would peruse in his bondage, plucking the fruit and for good measure the leaves. That was the kind of undiscriminating appetite Vernon had—bottomless, never letting a scrap, edible or not, go to waste.

As soon as we got home from work and untied him, he was gone. I don't blame him. In our long happy and often exasperating relationship with the canine family, we learn and forget and rediscover that every annoyance and inconvenience and disappointment begins on one side of that equation, with us. In their shortcomings, we find our own. We usually found him Dumpster diving behind the China King or raiding the half-eaten bowls of the less covetous dogs in the neighborhood.

One night we couldn't find him anywhere. It was getting dark, that time of night when you stop driving around and start to give up on dogs, hoping to find them curled up at the back door in the morning. Then out the car window a flash of fur appeared in someone's floodlit backyard. And there he was, standing before a large cage, stunned into a point, hypnotized. Then we saw it. Reaching out from inside the cage, the tiny, almost human hand of a raccoon was taking swipes at his nose, beckoning him closer.

As he grew up, this dog of ours began to resemble other things entirely. Jim Henson spent his early childhood along the sloughs and bayous of nearby Leland, where Deer Creek, an old tributary of the Mississippi, is claimed to be the birthplace of Kermit the Frog. Vernon favored several Muppets, Animal most prominently. And at the King Biscuit Blues Festival in Helena, Arkansas, he disappeared over the riverbank and emerged in the jack-o'-lantern twilight as fuzzy and green as Kermit. His protean nature had a Jekyll and Hyde aspect to it. Vernon was mostly a gentle and uncomplaining soul, but offered a heel of bread or a crawfish tail or the leftovers from one of those monumental T-bones at Doe's Eat Place in Greenville, he would devolve into a snarling, slavering beast and crawl under a car or inside his fig tree, growling and snapping if you came within ten feet. Get closer and he might take your hand off.

Leaving is as much a part of the Delta's checkered history as putting down roots in the miraculous soil. And like many Deltans before him, Vernon made his way out, first to Baltimore for my grad school, then to New York for Kate's grad school. It was a path dictated like those of other Delta expats by education and opportunity, and he took full advantage.

Up north they had never seen anything like Vernon before. In Central Park, where all varieties of dog high and low congregate to run off leash in the mornings, we could hear people guessing at his breed. At the dog run of our Harlem neighborhood, the children who lined up along the chain-link fence to watch the dogs would point to him and say, "Ooh, Mama, I want that dog!" Whenever anyone asked what he was, at first we tried to formulate the various breeds. "A little bit of everything," we would say, which after a while was boiled down to "whatever you want him to be."

There were comparisons to a Muppet, a Fraggle, a Creamsicle, a Scottish sea captain, and way too often to Falcor, that wise and long-suffering dragon-dog from *The NeverEnding Story*. One day, as I walked him out the front door, someone passing by called out, "Waddup, Chewbacca!"

Everywhere he went, he was an oddity and a wonder. I have walked him down Bayard Street, the main drag in Chinatown, and might as well have been leading a hog on a leash down Main Street. I have walked him down the stairs of a fifth-floor walk-up in Little Senegal on 116th Street and watched people cower in their doorways at the sight of him. The sidewalks of New York offer an all-you-can-eat buffet of chicken bones, pizza crusts, fried rice, and half-eaten bagels. One day on a walk down 122nd Street, a chicken wing fell from the sky, or at least five stories from the building above, which convinced us of his powers to perform selected miracles.

That place on 116th was our first home in the city, a building that basically operated as a Senegalese village. The apartment below ours was designated the village kitchen. The door would open, and people of all ages—men on breaks from their taxi shifts, women robed in a rainbow of fabrics and headdresses, entire families—would emerge carrying stacks of Styrofoam containers trailing a fog of stewed fish. I have eaten Senegalese food before. It melds an unlikely litany of ingredients, cultures, and continents, sour and savory in all the right ways. And after smelling it being cooked from a rank base of some kind of fermented sea creature, I can only describe it as alchemy. Every afternoon nauseous fumes would bloom up through our windows and floorboards, and Vernon, resigned by then to the boredom of apartment life, would lift his head up off the floor and start sniffing around, trying to guess at the source of that glorious aroma.

The families who lived in the building were nice enough to strangers who had invaded, if not quite infiltrated, their culture. But they still froze in terror whenever Vernon made an appearance. More than the structure of Senegalese society or the ingredients of Senegalese cooking, this was the greatest mystery of living there. We dreamed up all kinds of conspiracy theories. Maybe it had something to do with religion and the uncleanliness of dogs in Muslim cultures. Maybe in Senegal dogs had been used by colonial powers to intimidate or even torture people. Maybe it was just that they came from a similarly urban environment where people are smart enough not to keep anything that requires feeding. We had no clue.

Then one evening as Vernon and I were escorting Kate's sister to the subway, we passed a man in the hallway dressed in a long white tunic and speaking another language into a cell phone. We had reached the next landing when he broke off the conversation and called down to us. "Excuse me," he said, peering over the railing at Vernon. "Sorry to ask, but"–and here he swallowed as if unsure about asking–"is that a lion?" So that was it. Here was this guy, probably talking to folks back home in Dakar and telling them how crazy Americans are, keeping lions as pets.

In a way the guy was right, though. Vernon with his great blond mane and piercing russet eyes was as much a lion as a dog ever was. He also favored a gazelle, a wildebeest, and any number of other species a lion might prey upon. He was whatever you wanted him to be. And this, as much as calling food from the sky, was his particular miracle. Like Whitman, he contained multitudes. At least to us he did.

When a pet dies, as with any beloved person or thing, you do not just mourn the departed. You mourn the life you've lived along with the departed. In this case that was nearly a decade–only a decade–that brought our marriage and honeymoon, the birth of our first child and the expectation of our second, all the places moved into and out of, the deaths of family and strangely enough the deaths of every one of Vernon's brothers and sisters who had found homes at that party ten years ago. We always intended to bring Vernon back home, to return him to the country whence he sprang, where he could live the free and easy life a dog was meant to live. He never did make it.

But in looking back at his life, we can see our own in reverse, unreeling back down I-95 and across the back roads of the South to that backyard in Indianola, the house gone now, demolished, leaving a clear view of that sprawling fig tree. The tree under which we hope to lay his ashes to rest. We can see how far we've traveled, and how far away from home we've strayed.

CHAPTER 4

Family Ties

A Marriage for the Dogs

BY JILL McCORKLE

When my husband and I got married, we were as concerned about merging our dogs as we were our children. Turns out the children were remarkably easy, but the Brady Bunch thing is a much more difficult and complicated process when it involves teeth and ticks and peeing on what you think belongs to you. He had two border collies who had always lived outside and whose days were all about working hard. Rufus and Okra are siblings–Rufus, a handsome redhead, is so intelligent you expect him to open his mouth and speak in complete sentences, to quote Shakespeare or ask to borrow the truck. Okra is black and white and needy all over. She was the runt of their litter and clearly did not get enough attention from their mother. If she could talk, she would be saying, "Pet me, pet me, pet me, love me best." If Rufus could drive the truck, he could take her to therapy to work on some of these issues. Their visits inside the house were limited to freezing cold nights (a rare thing in North Carolina), and neither possessed cuddly stuffed toys. They ate whatever was available, and their feet were perpetually stained orange from the red clay lining the pond and the banks of the Eno River where they spent what little bit of leisure time they had attempting to catch frogs and fish and chasing deer. They are rustic back-to-nature dogs who live simply and work hard. This is the life they have always known.

Meanwhile, my not so neatly assembled pack were all from the North. Vanessa, a yellow Lab from Massachusetts, had spent much of her life lounging on the sofa watching television. She was so sweetly submissive that she couldn't help but pee a little upon greeting people—a problem for which I sought help from an animal behaviorist but with no success; she also had a little attachment problem where she had to have something in her mouth in order to go outside. If she couldn't find one of her stuffed toys, she would grab the nearest fabric she could find. At the end of one particularly snowy frozen winter, I marveled at how I seemed to be missing a lot of socks and underwear, only to discover with the thaw that Vanessa had left a good portion of my belongings in the driveway. She was the largest and the oldest in my pack but had no authority whatsoever, which didn't seem to bother her so long as meals were served and there was room on the couch.

At the top of the pack was Daisy, a sheltie who was a year younger than Vanessa. Born in Rhode Island to a breeder who also raised Jack Russells, Daisy had come into life a little hyper and nervous. She circled cars in the driveway and children in the yard. She rounded up Vanessa when she was moving too slowly. She was the best babysitter I ever had, and even when the kids were teenagers, I would hear the clicking of her nails as she tiptoed room to room to make sure they were where they were supposed to be before flopping down and finally closing her eyes. She was ever vigilant though still not above the stuffed toys and cookies Vanessa thrived on.

And last there was Buster, a Shih-poo-whatever mix who had been passed off to us as a papillon because that's what my son, then eight, had said he wanted. He had been asked at school to name what he would wish if given three wishes, and he had said: (1) to live a regular life plus a hundred years; (2) for everyone he loved to live a regular life plus a hundred years; and (3) a papillon. It was at the height of a time when he was asking all sorts of dark existential questions about death, and so granting that particular wish seemed an easy thing to do. And there was my pack of Yankee dogs–small, medium, and large–whom I now had to ready for a move below the Mason-Dixon and not just south but into a life with goats and chickens and stepdogs who had never spent a day of their lives sleeping on the couch and watching *Sex and the City*, as Vanessa had spent so many hours doing with my daughter.

The sixteen-hour trip loomed before us. No way to do it but a straight drive. The cat, Crystal, was in the way back in a crate, Vanessa stretched out on the backseat, Buster perched on the console like a little hood ornament, and Daisy–nervous about the whole event–on my lap, drooling and passing gas as she always did with difficult transitions. The vet had given me a pill to give the cat if she didn't settle down, and I was tempted to take it myself on that ninety-degree day, somewhere around the Bronx where my gas light came on while in bumper-to-bumper traffic on a bridge and

Vanessa got carsick. I saved the pill as a keepsake and reminder of how things did not get as bad as they could have.

When the merge happened, the Southern dogs were warm and friendly, wet with river water and happy to greet the new arrivals. Vanessa thought this was all fine and just wanted something to eat and of course to find her way to the nearest air-conditioned room and couch. Daisy—more finicky about her associations—spun in circles, round and round and round, and could not find a comfortable spot to flop. She busted out a screen on the porch and gnawed up the bathroom door. Buster made it perfectly clear that he doesn't like to share anything—not food or bed or toy or human; he has the Napoleon complex and always has to pee last and was anxiously marking everything in sight. Rufus and Okra became confused and forlorn as they stood with their noses pressed against the glass, not understanding why the new guys were allowed where they had never been. This is when Rufus—in another form—might have gotten the keys to the truck, said something cool like "Suit yourself," and driven around the back roads to try to make sense of it all, while Okra kept begging them to like her, to like the South, the humidity, and the snakes, the parade of possums and raccoons they would all soon encounter. "Like me. Like me. Like me. Like me."

It wasn't too long before Vanessa was swimming in the pond several times a day and Daisy had come to appreciate the shade of a big bush where she could watch with great vigilance the coming and going of any car in the driveway. They all came to appreciate the fine delicacy of goat and rabbit droppings and the occasional stray chicken. There were even times when Rufus (maybe he thought no one was watching) collected a few stuffed toys Vanessa left around the yard, and stretched out in the shade to take a nap. It was paradise, and I was happy that Vanessa and Daisy got to enjoy that brief window of Southern retirement before old age took them.

And this is when the Brady Bunch became Yours, Mine, and Ours. In dealing with the deaths of Vanessa and Daisy, we decided to get a puppy. Enter the big baby–a Bulgarian shepherd, bred to guard livestock. His given name is Znam, but we call him Zeno or just Z, and he wouldn't want to sleep inside if you offered it, preferring to stretch right out in a cold driving rain. He is one with the elements. However, he does love a cookie, and he does like toys, though his tend to be tools he pulls from the barn or dead squirrels or deer skulls he brings up from the river. We have often found him tossing and fetching a possum playing possum, and he handles snakes pretty regularly. He doesn't seem to mind a bit that Buster continues to come and go in and out of the house and pees after everyone else or that Rufus is in charge. He's a working dog, but he also loves to play and then stretch out and take long naps. He wouldn't go to the city if you paid him, but Rufus might, and of course, he would want to drive.

Dog Meets Wife

BY CHARLIE GEER

My wife doesn't come from dog people. In the Andalusian pueblo where we met, her hometown, most dogs exist to work–as herders of sheep, trackers of boar, defenders of real estate–and when they aren't on the job, they are simply fed, housed, and tolerated, like any other farm animal. They do not sleep on couches, ride shotgun, meet up for puppy playdates. They are not family.

My dogs always have been. I didn't have a pup at my side when I met Concha only because I'd recently lost mine to a vicious pancreatic tumor. Suttree had been with me for over a decade, an exceptionally devoted friend and cohort, and it felt far too soon to try to replace him. (It still does.) Months after he passed, I'd expect to hear him snuffling eagerly at the door when I came home, sixty pounds of spotted spaniel glee overjoyed to see me. When I talked to Concha of Sut's adventures and misadventures, recalled how heartening his presence had been, my composure would sometimes crack. Concha's consolation was generous but a little awkward. I was talking about a dog? Not a person? "Can't help it," I finally told her. "I'm dog people." She met this admission with her lost-in-translation look, face drawn into a puzzled, faintly skeptical squint–precisely her response when I'd tried to explain dry counties.

"Dog people," she said. "I do not understand this. It seems very strange to me."

"I know. It's just that where I'm from, dogs can be–"

"People?"

She didn't get it until she traveled with me to South Carolina to meet her future in-laws, and as a matter of course their dog, Ludwig, a stone-deaf Llewellin setter my mother rescued from the breeder's cut ten years ago. The pup nobody wanted–what good is a bird dog who can't hear a whistle?–had breezed merrily into the fold, an affable addition to a nest long since emptied of children. Whenever my thoughts turned homeward, Ludwig was there to round out the tableau. Even living on the other side of the Atlantic, I couldn't imagine life without him.

Concha could. When she first met Ludwig, she treated him like a piece of furniture, something to avoid bumping into or stepping on. She didn't refer to him by name, but simply as *el perro*, and she ignored his entreaties, perhaps not even recognizing a nose to the hand as a gesture of good faith. Ludwig learned quickly that if he wanted a chin scratch, an ear rub, or just a kindly dog-people vibe, he would have to look elsewhere. But tuning in to the daily rhythms of the house, Concha began to discern just how integral *el perro* was to domestic life–how familiar, in the Spanish sense of the word ("of the family"). Concha watched him accompany my mother as she went about her gardening and assorted home-improvement projects, witnessed the delight, on both sides, when the dog welcomed my father home from work. After a couple of days, Concha began to scrutinize Ludwig rather than shrug him off, as if she might be open to persuasion. On day three, I even caught her pat Ludwig on the head–speculatively, and only once, not a proper head scratch by any measure, but a step in the right direction.

The next morning she walked up on me talking to Ludwig and I figured I'd blown it, all that progress for naught. Topping off a bird feeder in the yard while Ludwig looked on, I was just chit-chatting away, the way dog people do. How must this have looked to someone who wasn't dog people? Here I was, a reasonably sensible human being, talking to an animal, and a deaf one at that. Worse, I wasn't simply talking to the deaf animal, but *querying* him: How had he been getting along? Couldn't he do something about that rotten seed-hoarding squirrel?

"You talk to him?"

Startled, I fumbled the feed bag. "What's that, sweetheart?" I said, as though I hadn't heard her question, didn't understand perfectly well its implications.

"You talk to Ludwin?" she said, softening the Teutonic handle with a Latinate twist.

"I know. It's kind of–" Hang on. Had Concha just referred to the dog by name? I looked up. Ludwig was nosing her hand, and she wasn't pulling away.

"It is like he is a *compañero*," she said, smiling down at the dog. "Ludwin is your *compañero*. No?"

compañero n. companion

"Sure," I said. "If you like."

"It does not matter if he hears you. He is here and you are here together. *Compañeros.*"

"Yes," I said. "Exactly. That's exactly it."

Concha put her free hand on my shoulder. "You and Suttree . . . ," she said. "You were *compañeros.*"

"Yes," I said. "Yes."

Ludwig made a model ambassador. You might reckon a deaf dog even needier than most, but just the reverse was true: Because he often didn't even know we were there, Concha and I could go about our business while he went about his. He is gentle by nature, unobtrusive, and unless he's been out tromping through pluff mud, his handsome tricolor coat is invariably bright, clean, and plush. He invites goodwill.

Disability might have denied Ludwig the stately pursuit of upland game, but from day one he has made do with backyard quarry–squirrels, pigeons, grackles. You can take the bird dog out of the hunt, but good luck taking the hunt out of the bird dog. Afternoons on the porch Concha and I watched him point and flush. However inelegant his prey, the quest itself is a thing of beauty. On point he locks in, a tightly wound spring: tail stiff, body rigid, weight canted forward. Doing what nature wired him to do, he is a picture of precision and control. Curiously, his ears figure in every point. As radar they are useless, but still they cock forward, shiver with anticipation.

"Ludwin!" Concha squealed one afternoon after he broke point and the birds scattered. "He is amazing!"

It's true Concha never quit looking a little alarmed when Ludwig hopped up into my father's lap of an evening, but by the end of the week she had come to understand why a dog might do that, and why you might let it. She had come to understand how this worked, what the two species were getting out of the deal, just how wondrous unconditional companionship can be.

I knew her conversion was complete when, toward the end of our stay, I caught her talking to Ludwig. From down the hall I could see the two of them, Concha folding clothes for the suitcase, Ludwig seated on the floor watching her. I couldn't make out what Concha was saying, exactly, only that it was Spanish and that it was tender, playful, and occasionally inflected with the interrogative. I didn't say anything, didn't let her know I heard her. But later I gave Ludwig a good chin scratch, piled on praise he could not hear but could surely sense.

It may be a while yet before we get a dog of our own. Even as Concha is learning how to let dogs in, I'm learning how to let one go. But I have a feeling it will happen one day, and when it does, I know just whom to thank.

Golden Snapper

BY LISA HOWORTH

Oh, Puppy Sal was a snappy old gal,
it was agreed she was one in a million.
She loved most folks and dissed most dogs–
she didn't know she was one.

FROM "THE BALLAD OF PUPPY SAL" (TRADITIONAL)

When my husband and I first moved in together, we came with baggage–the furry kind. Mine was two cats, and his was Puppy Sal. Richard came from a no-cat-zone family; I'd never had a dog. I had serious reservations–a *bad* dog chomped my face when I was a baby, and I have issues with things like road-kill rolling, underwear chewing, and litter-box snacking. Puppy Sal did all these things (face chomping excepted) and more that are unmentionable here. It was impossible not to immediately adore this copper-colored mutt puppy who appeared–accreted in–from the woods around William Faulkner's house and followed Richard home. Puppy Sal was, or was going to be, a smallish-to-medium-size dog with a gorgeous blond plume of tail. She looked like a miniature cross between an Irish setter and a golden retriever. A little.

I myself accreted into Oxford, Mississippi, at about the same time. Right away we took Puppy Sal to the vet for shots, but apparently she'd already contracted distemper and became so gravely ill she had a grand mal seizure. We thought she'd have to be put down. The old-school vet, Dr. Harland, believed he could pull her through, although he warned, "She might exhibit some strange behavior."

Puppy Sal did pull through, and began snapping her jaw, once a second, in perfect "one-Mississippi, two-Mississippi" measure, like a heartbeat, even in her sleep. This distressed people, but not Sal, who snapped those golden mandibles for the rest of her fifteen years. The only time the affliction ever handicapped her was once when she got hold of a big cat-head biscuit from the Jitney Jungle parking lot. She came home whining, the hockey-puck-like thing firmly lodged in her whopper-jawed mouth.

Puppy Sal knew lots of tricks; a favorite with her many friends around town was falling to the ground "dead" when you cocked your hand like a pistol, pointed it at her, and said, "*Bang!*" She would remain dead, waiting to be brought back to life by the words "Good dog!" and a vigorous belly scratch. She was a famous chanteuse, her favorite song being the Righteous Brothers' "You've Lost That Lovin' Feelin'," which, when we sang it in a high falsetto, would inspire her to throw back her head and croon along with us, as we began, "You never close your eyes anymore when I kiss your li-i-ips." These accomplishments and her happy personality earned her many fans, who learned to accept the unfortunate snapping. In fact, it only enhanced her celebrity. When she traveled the two blocks to the Square—this was most days—you might see people requesting a performance. Puppy Sal knew no strangers, and if an office or a shop had a door open, she was in there, checking on things.

She made a few enemies. Mr. Paisley, a fastidious gentleman with a lisp who was the manager of the Holiday Inn, called regularly to complain that "your dog" had been hoovering up treats that had escaped from his Dumpster, and then, her ears slicked back with steak or catfish grease, had gone for a swim in the hotel pool. There she'd lie on the top step, her jaws at water level, snapping a drink and leaving "a dithguthting earl thlick" on the pool surface. Sal was often apprehended by James Burt, the friendliest dogcatcher who ever lived, who not too secretly had a crush on her and would just bring her on home. Oxford was small and slow then and had no leash law, or it wasn't enforced if there was one; dogs could be dogs; dogs strolled around town, dogs slept in the street. We were irresponsible, I regret to say, but we were young in Puppy Sal's lifetime.

Puppy Sal preferred human company, was afraid of cats, and scorned other dogs, but she did have a suitor whose companionship she once permitted: a dashing silver husky we called Mr. Sweetheart. He'd come courting, dragging a chain and a stake. He and Sal were caught in flagrante delicto in the old cemetery, where, for lack of a more delicate way to put it, they had difficulty disengaging. Or as our friend Larry Brown described the situation, "They done got hung up," requiring assistance. The result of this one-night stand was a litter of five puppies, none of whom looked like either parent, or one another. A fair mother, Sal would stand in the pup pen, surrounded by her mewling sons and daughters, looking confused and indignant, as if to say "What *are* these things and where is my nekkid, pink human baby?" To find the puppies homes, we placed a classified ad in the *Oxford Eagle* with the heading "FREE: ONE-HALF GOLDEN SNAPPER PUPPIES." Most takers got the joke, except the first to arrive, an excited foreign graduate student couple who spoke little English. They were disappointed at the obvious Heinz 57 selection, finally asking, "What *is* 'Golden Snapper'?" They went away without a puppy, no doubt complaining to each other about the problem in America of false advertising.

We lived for two years in Washington, D.C., where Puppy Sal adapted well to city life and the elevated status of being walked on a leash. She quickly cultivated a new cadre of fans around Thirty-Fourth and P Streets in Georgetown, close to the old Savile Book Shop, where we worked. She walked to work with us every day, and then assisted the German bookkeeper upstairs, where it was quiet and the air conditioner and the radiator were in good order, not like in the frigid or stuffy decrepit old shop downstairs. Sal had a preternatural sense about Accounts Payable, the bookkeeper insisted one day when "Der Pup" stood at the top of the office stairs, nose wrinkled, lips curled, baring those snapping choppers at a certain diplomat whose embassy account was in arrears. Or it could have been something else about him that alerted her instincts. The man did not venture up, although Puppy Sal never bit a soul, and probably couldn't anyway. She saw herself as a belle of the rural Deep South, celebrated and special in the big city, winsome and gracious. I can remember her bowing up only one other time there, when a pleasant British gent came around a corner with a dog that was damn near a clone of Puppy Sal. A

Golden Unsnapper. Sal froze, giving the faux Sal the once-over before displaying a haughty snarl of teeth and gum. It was as if two women–say Pamela Harriman and Susan Mary Alsop–had encountered each other at some Georgetown power soiree wearing the same designer dress.

When we returned to Oxford, a new leash law, among other refinements, had been instituted, which we feared would cramp Puppy Sal's old small-town style–her daily freewheeling constitutionals. People had missed her and were afraid they'd never see her tricks again, but Sal did not disappoint. Her time in Georgetown had taught her that if she was going to show out, she was going to have to do it on a leash. Sal had become a shining example of canine comportment, demonstrating a sophisticated understanding of citified leash-law protocol, still able to do her tricks, albeit somewhat impeded, back to being a Big Pup in a Small Town. We thought of her as a sort of Eliza Doolittle, My Fair Puppy, although I'm sure she would have preferred to be likened to My Fair *Lady*.

Did Puppy Sal ever miss city life? There was one event that made us wonder. We'd gone with friends to swim, drink, and throw bottle rockets at each other at Sardis Lake (where Faulkner sailed his boat, the *Minmagary)*, as we often did in the brutally hot summers when there was little else to do in Oxford. Other dogs were along, and they happily ran and splashed about together, performing nice tricks like *fetch*. Puppy Sal was having none of that doggy lameness, and sat apart on a clay bank, disdainful and aloof. At some point, unbeknownst to us, she plunged in the water and began swimming–person-paddling–north across the lake, where some boaters eventually spotted her and brought her back. Affronted that she'd been expected to consort with other dogs, did she spy the dogless party boat way out in the lake and decide that that was the company she preferred? Was she striking out for D.C.? We'll never know.

I'm certain Our Fair Puppy expected to go to People Heaven and not to be simply relegated to Dog Heaven. As far as I'm concerned, if there really is any kind of People Heaven at all, dogs will definitely be there, too. I hope Puppy Sal is dealing with it in her most charming, ladylike way.

The Dog Who Lived Forever

BY NIC BROWN

Just before her second birthday, my daughter began requesting bedtime stories. My wife and I had always read books to her, but now she wanted tales of my own creation. I'm a writer. I can do this, I thought. Over the next few nights, I gamely generated a good one about a kitten lost in the snow, then another about our neighbor's dog sailing a boat. But soon I started drawing blanks. Finally one night, a couple of days later, at a loss beside my daughter's crib, I said, "Once upon a time there was a dog named Annie." And that's when it all began.

You get only one first dog. Yes, you can have dogs all your life, but those that share your childhood loom with extra import over all subsequent years. My very first dog, a sheepdog named Shadrack, was old when I was born and died by the time I was three. I have no memories of him, so he doesn't really count. But Annie, Annie was *the* dog, the dog of my youth.

The first Annie story I told my daughter was the creation myth. It goes like this.

In 1982 I was a five-year-old kindergartner in Greensboro, North Carolina. One lunch period an emaciated black-and-white puppy loped out of some scrub pines and into the playground. Apparently constructed out of little more than bones and questionable border collie genes, the dog knew she had just struck gold. (Gold, in this case, meant a bounty of peanut butter and jelly sandwiches provided by me and my friend Ralph.) We fed her for about a week before my mother caught wind of the situation and put the dog in the car with the rest of us when she picked up carpool one afternoon. The thing was so sickly that before we even got home, Mom stopped at the vet and left her with Dr. Ken Eiler, ostensibly with the idea that he would put the emaciated dog to sleep.

Three days later, the vet called the house.

"Pam," he said, "come get your puppy. She's firing on all six cylinders now."

What was a mom to do? Against all odds the creature was still alive, coaxed back to life by the miraculous Dr. Eiler. So we picked her up. If a starving animal can look brand-new, this dog did. Her fur suddenly shone and smelled of clinical shampoo, but she was still lanky and bony, just a cleaned-up hungry puppy. She was a gentle girl, though, polite even in her need. My brother announced, "This is the best free dog we ever had!" not knowing that Mom had just footed a $350 vet bill. Mom suggested we name the dog after Little Orphan Annie. And so it was.

At home we filled Shadrack's old dog food bowl with kibble, and Annie ate until she fell asleep with her head in the bowl itself, as if guarding the meal at all costs, not sure she'd ever have another.

I guessed my daughter would cotton to the image of a dog asleep in a food bowl, and I was right. After I finished the story, she requested it again. She requested it the next night. Soon she wanted more, and so I told others.

I told her about how, one day, when driving home in our huge silver Dodge Prospector, my mother took a sharp turn onto Country Club Drive, and Annie, who was sitting in the passenger seat enjoying some fresh air across the tongue, fell out the open window. Mom didn't even notice until a few houses later when she looked over, found the seat empty, then turned to discover Annie stunned in the street behind us. I told her how, for my school's annual fall festival, I dressed Annie as a Mexican bandito and entered her in the Halloween costume contest. I told of the Christmas chocolate incident of 1985, of Annie's predilection for chewing felt-tip pens into shreds on the dining room carpet, of walks around the neighborhood, of tennis ball catch, of countless dog minutiae.

The requests for stories became constant. "Tell me an Annie 'tory!" my daughter would chirp, the *s* disappearing somewhere inside her tiny mouth.

I told all I knew, but after a month or so, I started to run out of memories. Though Annie was always part of my daily existence, like some benevolent childhood atmosphere, not every day of my childhood generated a story worth remembering, let alone retelling. I recalled the coarse skin on Annie's back where she always chewed off her hair, her black gums, those dark splotches laid out across her belly like islands on a map, but it's hard work turning a splotchy belly and a dog's bald spot into good bedtime stories. So I started making them up.

First it was Annie entering our neighbor's house in an effort to steal food. Then Annie's friends, Happy and Molly, began to appear. Soon they all started to talk. The stories became wild, ridiculous, filled with magic and neighbors.

"Annie 'tory!" my daughter would say, every night, and off I'd go.

Requests soon left the bedroom and entered the rest of the house, coming at all times of the day. For the first time in years I found myself thinking about Annie constantly. I don't have a sophisticated belief system about afterlife, but one night while washing dishes I had the epiphany that, at that moment, Annie was one of the most important things in my daughter's life. It was a resurrection, a piece of my childhood I was magically sharing with her.

This went on, I kid you not, for almost two years straight. That's half of my daughter's life thus far. But recently, her obsession with Annie began to subside. Requests receded once more to the bedroom just before sleep. And then, a few weeks back, she asked for a story about her doll instead. Despite the fact that I had so often rolled my eyes at the requests for Annie stories and longed for them to cease, I now found myself hoping my daughter would ask for another. But she did not. I felt like Annie had died a second death.

I grew adept at telling doll stories in place of Annie ones, and they put my daughter to sleep just as well. But the other night the pattern changed again. We'd kept her up too late (9:00 p.m.!), and so, as any parent knows, an emotional meltdown ensued. I read books to her, told a doll story, but all to no avail. There was nothing that would soothe her. In tears, overwhelmed with fatigue, she finally sighed and said, "Tell me an Annie 'tory." And suddenly Annie was back, running through the streets again, talking to owls, swimming with mermaids, flying through a thunderstorm. My daughter fell silent, comforted by the image of that other love of my life, eyes closing, and gone.

We don't have a dog in our family these days, but my wife just told me that the shelter in Oxford, Mississippi, where we moved only weeks ago, houses more dogs than any other town's in which we've lived. Maybe it's time we went down there and adopted one, if only so my daughter can generate her own stories someday, when, at a loss beside a crib one night, she finds herself mining memories for tales to lull her child to sleep. Because she's already learned that when you find the right story, one about something you really love, people want to listen. And sometimes, when the story is about the thing you loved most, the people listening close their eyes and start to dream.

Two of a Kind

BY BETH HATCHER

Protectors are often misunderstood. Big-Dog was, and so was my aunt.

Both the bulldog mix and my dad's older sister possessed a crackling fierceness others often mistook for meanness. And I did too, once, until observation and age showed me that both aunt and dog were a little softer than they let on, at least to those they truly loved–to those they wanted to protect.

I can't help but think of my aunt Betty Joe whenever I think of Big-Dog. The pair were always together and in my memories have sort of fused into a single mythic entity. And if you don't think an aging old maid has anything in common with a massive brown-and-white bulldog, you never met my aunt.

Aunt Betty Joe was an old woman by the time I was young, a wrinkle-crossed, chain-smoking, tough-talking high school teacher who'd worked her way out of the tobacco fields and through a master's degree in chemistry. In the 1940s North Carolina of her childhood, most kids went straight to the farm, especially a sharecropper's daughter, but Betty Joe had other ideas and she wasn't afraid to fight.

Neither was Big-Dog.

Big-Dog was young when I met him. Family lore has it that Betty Joe chose him as a puppy at the pound after seeing him fight a much bigger dog that had taken his food. "That'un there," she'd said, a cigarette probably dangling from her mouth. "I want that one."

She didn't want a lapdog, after all. She wanted a big tough dog that could stand guard over her sandy country acres, where copperheads and petty country criminals sometimes slithered through the trees.

So she brought Big-Dog home. His arrival terrified most of the extended family, especially me. I lived close by and was just a kindergartner at the time, so my first memories of Big-Dog are fuzzy, mostly him snoring on my aunt's screened-in porch. I do acutely remember his size. Even as a puppy he was huge—and oddly shaped, since the American bulldog (and whatever else) in him made his legs long and lanky, while his massive square head seemed suited to a much stockier body.

The awkwardness of his frame as well as a drooling, lopsided underbite only contributed to his terrifying demeanor.

He was graceful when he ran, though, almost feline-like in his quick movement. Watching him run down a snake or a ball was like watching a strange-looking cheetah.

His eyes were like a cat's as well, not in their shape but in their gaze. Big-Dog never looked at you with that eager neediness of most dogs. He'd always look at you with a sort of patient disdain, the type of gaze that said he was never going to do anything that *he* didn't want to do.

My aunt had a gaze like that as well. She didn't do anything that she didn't want to do either, especially waste time acting like a "proper" Southern lady. She wore fishing hats full-time, made off-color jokes at the general store, and stayed as far away from marriage as most people did from Big-Dog. I loved her, but she could be a little bristly. Sometimes she scared me even more than Big-Dog. At least the bulldog couldn't eviscerate me with a few well-chosen words.

My mother would say that my aunt's hard life had sharpened her tongue and that her "bark was worse than her bite."

Big-Dog's wasn't.

His bite turned out to be exactly as bad as his bark, which a neighbor lady found out when she came calling on my aunt one day. My aunt told the woman to stay in her car until she could shut Big-Dog in his pen, but the lady said she had a "special way with animals" and got out of her car anyway. Big-Dog went for her immediately, or "just put his teeth just lightly around the woman's elbow," according to my aunt.

The county called it "biting," and this, combined with a few other of Big-Dog's protective instances, got Big-Dog sent to the pound. My aunt's heart broke, and for the first time ever I saw tears in her eyes after Big-Dog had to leave. "He was just trying to protect me," she said. "They don't understand." I cried for Big-Dog too, because in the five years since my aunt had brought him home, he'd become my trusted friend as I'd begun to see his softer side.

We spent hours together in the woods between my aunt's house and mine. In this secret piney world Big-Dog and I ruled supreme. He might be a dragon, or the prince or whatever make-believe role I could dream up that day. But as long as he was with me, my mother never worried about me staying gone for hours. She knew he would protect me, and she was right. Once he saved me from a copperhead snake and another time alerted my dad when I twisted my ankle over a root. In quieter hours he let me paint his toenails and listened to all my problems, his great big drooling head lying sweetly across my knee. Tender and tough at once.

In fact Big-Dog was so tough that he broke out of the pound soon after being incarcerated and somehow made it back to my aunt's house across several swampy miles. My aunt thought Big-Dog had learned his lesson, and I'm not sure how she worked it, but he never went back to the pound. Big-Dog stayed with her the rest of his life, right by her side and living like a king. He protected all of us until the end—even scaring off a burglar in his last year of life.

When Big-Dog died, I was just entering high school, and my aunt was retiring from it. She hung up her teaching hat as a chronic disease began taking the last of its toll. During these final years of her life, I finally saw *her* softer side. And I began learning about all the things she'd done for the family that I hadn't known about.

I hadn't known that she'd come home as a young woman to care for her ailing parents and help out her ten brothers and sisters. I hadn't known how many of my older cousins she'd helped raise and how often she'd stepped in to voice concern for them. I didn't know how much she'd helped me behind the scenes during a tough time in my own adolescence. I didn't know that the disease she suffered from had affected her since her twenties but she never had time to complain because she was too busy helping everyone else–teaching students, running a Sunday school class, and being right quick to tell anyone where to go when they spoke the wrong way about her family.

I kept learning these things even after my aunt passed away, her death leaving a hole in my family that's never been filled. I think only now as an adult can I truly appreciate why my aunt loved that big ol' lopsided dog so much.

Big-Dog protected her the way she protected others.

And everybody needs protecting.

A Puppy and a Patch of Grass

BY JON MEACHAM

About two years ago, my wife and I made a slightly impulsive decision to move from New York City, where we had lived happily for close to two decades, to Nashville. Though announcing this to our New York friends didn't provoke the same reaction as, say, reporting that we were divorcing or that one of us had been stricken by disease, I do suspect we'd have gotten similar looks of mystification, puzzlement, and pity had our news been of a more traditionally calamitous nature.

It's a familiar leaving–New York ritual. One friend, a woman of Southern roots long resident in the North, wondered how we could give up a life where any problem could be solved by three magic words: "Call the super." The question people asked with the greatest frequency and bewilderment was simple yet profound, and certainly existential: *Why?*

"Grass and dogs," we'd reply.

Yes, our answer was glib, but as Henry Kissinger used to say, it also had the virtue of being true. I grew up in Chattanooga, my wife in Mississippi, in both the Delta and Jackson. We have three children, all of whom, at that time, were under the age of ten. When our first child was born, we bought a house in Sewanee, Tennessee, the tiny Episcopal college town William Alexander Percy wisely christened Arcadia. Our wishful thinking was that ten months on the Upper East Side and summers on the Cumberland Plateau would be the perfect combination of differing cultures and climate for children who were Southern by blood if not birth.

As the years passed, however, it grew ever more difficult to pack up and re-migrate to Madison Avenue at the end of August. The brave new world of acreage, the freedom of being able to go out the front door and into something as radical as a yard, led us both to surreptitious online real-estate searches of suburban New York. It occurred to us, finally, that instead of trying to replicate the South we could just return to the real thing. Nashville seemed (and has proved) the perfect place.

The "grass" part of the equation was handled with some dispatch. A Sewanee classmate of mine from Demopolis, Alabama, had a brilliant friend in real estate in Nashville (auspiciously, the only other woman I've ever met named Keith besides my wife). We bought the first house she showed us, a lovely if faded Georgian Revival with a glorious yard and a creek at the foot of a sloping hill. So we'd checked the first box.

Now we needed the dog. As I think back on it, I believe there was more internal debate within the family over what kind of puppy we might get than there had been over the whole move itself. I had grown up with a liver-and-white English springer spaniel named Daphne, a brilliant hunter who merrily terrorized the quail on my grandfather's farm near Blythe Ferry on the Tennessee River. To me the choice seemed obvious, even stipulated.

I had momentarily forgotten, however, that my son and I live in a family of women. My wife wanted a Lab of some kind; my older daughter, Mary, wanted a beagle, having been struck years earlier by a *coup de foudre* on meeting the Platonic ideal of the breed in Henry, her godmother Julia Reed's exceptional dog; and the baby, then nearly four, was in solidarity with the sisterhood. Whatever the boys want, we want something else.

In domestic matters my son and I have a perennial political problem. There are three of them and only two of us. And they're louder. I was working on a biography of Thomas Jefferson when said baby girl was born. Our son, Sam, then seven, when told that he was going to have another sister, thus giving the female caucus a majority, replied presciently, "But Daddy, that means we're outnumbered." This was the same reaction the Federalists had when Jefferson bought Louisiana, provoking a secessionist movement led by New Englanders worried, rightly, about the expanding power of the South and the West.

In a moment of historical identification, I decided that Jefferson's strategy in the Louisiana crisis–to dispense with what he called "metaphysical subtleties" of constitutional law and move ahead unilaterally–had much to recommend it in our own unfolding canine crisis. Jefferson had created his own facts on the ground by pressing ahead with the purchase. So would I.

I called a friend in Nashville, a physician who had already made me feel very inadequate by asking me, before the deal on the house was closed, whether I had lined up a dentist yet, and asked him to hook me up with the local springer spaniel mafia. I figured there had to be one; what I had not figured on was Jim Sawyers, a self-described "dogman" based in Goodlettsville, just outside Nashville. Jim is kind of a dog NSA–he knows everything–and within hours, it seemed, he had found a liver-and-white English springer puppy in southern Ohio.

Within moments–and of this I am certain–of Ellie's arrival, the acquisition of a springer had become everyone's idea. (The same thing happened with Louisiana, by the way.) Beautifully marked, boundlessly energetic, and, it would turn out, sweetly theatrical, Ellie transformed the household. As she grew out of the cute waddling puppy stage into a lean early adolescence, she turned the yard into her magical kingdom. With balletic leaps out the door and over the steps, Ellie became a Belle Meade huntress, giving tireless chase to what our younger daughter, Maggie, calls "burrrds."

Ellie's daily hunting reminds us of Samuel Johnson's remark about second marriages–they're the triumph of hope over experience. We don't think she has actually ever caught anything, save for one already dead rat she pitched up with one afternoon, but Lord, does she love the hunt. From my desk I can watch her spend a good hour or more in wearying pursuit of game in an open rectangle of boxwoods. I think the birds are taunting her, but I like to think they're doing so somewhat affectionately. When the family of possums that live in the underbrush between our property and our neighbors' amble by in the twilight, Ellie puts on a good show of scaring them. She doesn't scare them at all, of course, but she likes to think she does.

She has a Shakespearean streak, a sense that all the world's a stage. Prideful, she loves being watched as she bounds from one end of the yard to another. Galloping and leaping and preening and pointing and panting, she is incredibly fast, pausing only to try to climb tree trunks or to lift a front paw in what we're certain she thinks is a flattering pose. (And she's right about that.)

To think of her as a creature of the wild, however, would stretch things rather far. While Ellie adores digging for moles, jumping the creek, and pointing Southwest jets that are climbing out of Nashville airspace, she has also yet to find a bed in the house on which she is uncomfortable and finds it totally natural to join guests on the couch for drinks. She's puzzled when she's asked to leave–or, more accurately, when she's dragged off. Keenly able to manipulate her expressions, she manages to convey a more-in-sorrow-than-in-anger look of amazement that we would be so outrageous as to want some time without her licking a visitor.

She loves her house and her outdoor domain, and we like to think that she loves us. I won't pretend to have any insights about the inner lives of dogs. But I do know this. When the door swings open every morning and a liberated Ellie leaps into the yard, seeking prey–whether "burrrds" or a jetliner–she does so with joy, springing out into the world, delighted to be free again. As are we.

Coveting Thy Neighbor's Dog

BY BETH MACY

The dog made us do it. Under a blazing September sun in front of two hundred friends and relatives, my husband and I got married in our backyard on the rural outskirts of Salem, Virginia. Scooter, our beagle-collie mutt, had set us on the matrimonial course. But even a country dog had the sense to get in the shade in the 98-degree heat–beneath the caterer's pig smoker.

Scooter had a knack for jamming himself into the juiciest of spots. He'd first shown up scrounging for table scraps during a backyard party the year before. We learned later that he belonged to the family across the street.

But they didn't pay much attention to him, judging from the ticks my husband pulled from his ears, the way he ran loose throughout the little mountain holler, and his unquenchable appetite. He was always nudging himself under the picnic table or next to the grill.

Getting a dog had been our whole reason for moving in together. Too dense to admit we were falling in love, Tom and I decided that since we both wanted a dog, we should just rent a place together and get one. We found a funky brick ranch at the base of Fort Lewis Mountain with so much land that it took three days to mow the lawn. It had stone retainer walls and a root cellar that was cloaked in poison ivy, which meant that before long we were too.

Having grown up city kids, we were new to this country enclave, full of shotguns and wall-mounted deer, cars on blocks and roaming, average-looking chickens–back then no one gave a thought to raising Rhode Island Reds. We tried to run over a copperhead in our driveway once with a Volkswagen Jetta, instead of just hitting it with a hoe, if that tells you anything.

The road was named Texas Hollow, but locals have long referred to our section of the county as West of the Brickyard, a nod to the brick factory near Salem's western edge. When I told a native-born coworker we were getting hitched, he said we had it all backward. "The only marriage recognized West of the Brickyard is common law."

Scooter was part and parcel of the WOTB landscape, an almost-stray who wandered loose, his right rear leg jutting out when he ran–from the time he got too close to a hot rod roaring down the street. And he was a master cuddler.

Then the For Sale sign went up in the neighbors' front yard. I was crushed. We'd been trying for months to domesticate Scooter, extracting the ticks, letting him sleep in our house. He had become a shared hound, and, though we hadn't spoken of the arrangement with the neighbors, they had three other dogs and didn't seem to mind. By the time I sent my husband over to ask if we could take the dog off their hands when they moved, they nodded yeah, sure, and that was that. You might want to get him a rabies shot, they said, and we did.

They called him Bear, which must have been ironic, because he was the friendliest, easiest-going dog in the world. Didn't make a mess, didn't shed. And he seemed invincible, never once requiring a sick visit to the vet. He once consumed a giant chocolate bar I'd inadvertently left out on the coffee table, foil and all, and didn't puke.

We called Bear's people the Bumpuses, after the neighbors in the movie *A Christmas Story*. And for a while everything was great. Scooter/Bear was back and forth between the two houses, double-dipping on food. In preparation for becoming solo dog parents, we started taking him on walks, eliciting grins from the Bumpuses when we walked past with Scooter tugging madly on the leash.

A few weeks later, we came home from work and found the Bumpuses gone, their house emptied. They'd left no note, no forwarding address. No Bear. When the new neighbors moved in, we asked where the old ones had gone. They moved into town, they told us, on Bowman Avenue.

My husband looked at me, my eyes already ablaze with a plan, and shook his head.

"No," he said. "You can't just steal somebody's dog."

"Why not?" I cried. "They already gave him to us!"

"Well, for one thing, they've got guns."

I didn't have an iPhone or a GPS device in 1992, so I did what any newspaper reporter would do. I scoured the Roanoke Valley map on our newsroom wall and made a plan. I'd leave work early, drive down the back alleys of Bowman Avenue, and peer into every backyard, praying I wouldn't become a crime brief in my own newspaper.

Minutes into the recon mission, I spotted moving boxes and knew the fluffy mottled tail had to be close. The dogs were all there, inside a chain-link fence. It was late afternoon, so the Bumpuses were probably at work.

I looked around to see if the neighbors were watching. They weren't.

In my memory, Bear became Scooter the moment he jumped into my loving arms and took his place in my passenger seat. Lucinda Williams could have written a song about us. I drove four miles to our house in the country, across a river and a four-lane highway and, finally, down the meandering rural road that landed me and the wonder dog West of the Brickyard gloriously home.

It was two weeks before the doorbell rang. When my husband opened the door, Scooter shot out into the front yard. Mrs. Bumpus was standing there, shaking her head.

"We heard he found his way back," she said.

"Yep, he just showed up," my husband explained, which was not untrue.

She apologized for taking him without talking to us but said her husband had changed his mind as they were packing up. She reckoned if she took him now he'd probably just keep finding his way back again–across the river, the ravines, and the four-lane highway.

"Probably so," my husband said, which was not untrue.

We had full possession of Scooter for fifteen years, through three houses and two kids; through the usual boredom and heartache, and the unusual bliss. Not only did he encourage the matrimonial leap, he also nudged us into buying our first house–thirty minutes away in Roanoke. It seemed he'd been helping himself to another neighbor's chickens and, as that neighbor put it, "If I catch him again on my property, I'm going to drop him."

But how did he know Scooter killed the chickens and not some other dog? I asked.

"Saw him with feathers in his mouth."

We called a real-estate agent the next day.

So the country dog became a city dog and, before too long, a big brother too. Scooter accompanied my husband on 3:00 a.m. walks when the baby got croup and cold air was prescribed. When the second boy came along, Scooter's seat in the family hierarchy moved even farther to the back of the minivan, but he never complained.

He was a country dog who never learned to walk a mountain trail off leash (without bolting), never lost his frenzied appetite even after he got fat. Six hours before he died, Scooter had his last supper. It was leftover take-out pizza, and he'd scored it by leaping atop our dining room table–at the age of sixteen. He even got the cheesy bits off the lid.

He had the decency to take his last labored breaths on a night when we'd been discussing whether or not to take him to the vet to put him out of his misery. He died at 3:00 a.m. while my husband slept on the couch, one hand in Scooter's fur to calm him as he panted.

Our latest rescue project is totally legit, I swear. Just before our firstborn left for college, he talked us into getting an SPCA puppy named Charley, a part Belgian Malinois resembling the Navy

SEALs' überdog, Cairo–the one who helped bring Bin Laden down. It would take SEAL training to break this high-maintenance beast of his shoe chewing, and a host of other bad-puppy habits. Unlike our old WOTB pal, this dog requires obedience classes and $500 vet bills and–the damnedest thing of all–actual playdates with other dogs. Every time I see him splayed on our sofa with legs askew, I think of the Bumpuses, who would surely smile if they saw our fierce Bin Laden hunter with his $60 haircut–and realize they're getting the last laugh.

If Charley wasn't microchipped, I might even find my way back to Bowman Avenue and leave them a replacement for the good dog that got away.

Home Is Where the Mutt Is

BY ASHLEY WARLICK

The last time I cried over dinner with my now ex-husband, we were in a dark little tavern in South America, drinking wine and eating shellfish, late at night. During our seventeen-year marriage, almost half my life we were together, there had been more than a few restaurants cried in: I'd told him I was pregnant in a restaurant, stormed out after a fight on our second anniversary, wept hormonally in a bar so long it became funny to both of us. But the last time I ever cried over dinner with him was about whether or not we could get a dog.

We were on vacation, for godssakes, and without even reliable Internet for looking at pictures of puppies, never mind near a shelter or a begging child. We were alone; we might have talked of anything. But there was plenty of stuff in the ether I had only begun to sense the edges of: I had just finished a draft of a novel that would need to go back to the drawing board; family was ill, aging, broke; and our marriage was on the cusp of its last downward spiral. I brought up the idea that we might get the children a dog, and the next thing I knew, the conversation had launched into old, dark territory.

Our previous dog had been my idea to begin with, too, and things had not gone especially well. I had insisted on her at the beginning of our marriage, and in her long, odd, unstable life, she'd kind of turned the man off animals altogether. (There's a metaphor here, but I'll not plumb to find it. Suffice it to say, when we put her down fourteen years later, her file at the vet was covered with stickers that read "will bite," although she never had.)

The idea of a dog became a kind of anti-talisman in the months ahead, a thing we held on to as we broke apart. Talking to the children about the fact that they would soon have two homes, both I and my ex-husband promised that now we could get a dog: this new we, the children and me. It was a small bright thing to offer, like bunk beds or your own bathroom, something to sweeten the pot. And a puppy: Nothing's sweeter than that.

The truth, of course, is that the balm is temporary. The puppy itself is temporary. It grows up. And here were these kids, these perfect part-people we had done a damn fine job of raising to ages nine and fifteen, and now we'd hurled them straight into the most adult of circumstances: a delicate bifurcation of home. With it, we said they could have a part-time, four-legged friend. What we were really offering was another life, in every possible sense of the word. It would start small. The question keeping me up nights was: What would it become?

The kids and I drove from our home in Greenville, South Carolina, to a little rescue outside Atlanta to get Frances after we saw her picture on Petfinder. She was nine weeks old, one of twelve or thirteen, reportedly part Australian shepherd, part Saint Bernard, liver spotted and white with a peppered snout and pretty pale eyelashes. Of all the littermates wrestling in their enclosure, she was the one I wanted. I couldn't really say why I was so certain, other than a kind of gut-level recognition I still trusted, in spite of the failures that had come before. I convinced the kids she was the one they wanted too, which was not hard because puppies are all cute in kind of interchangeable ways, and I honestly think they were having trouble telling one cuteness from another.

We named her Frances after Frances "Baby" Houseman in *Dirty Dancing*, the all-time best swoon of a movie we'd watched a couple of nights before, after running through a laundry list of silver screen possibilities: Brigitte and Audrey and Rita and Jean. She looked like a Frances, I thought, and these things, too, have a way of snapping into place.

On the drive home, the dog stretched across the laps of both my son and my niece; there was the sense of having been here before. There was a long road ahead, and not just to Greenville. I wish I could say that I was circumspect, that I remembered the bad dog with her sticker-covered file, that I thought about whatever might have gone wrong then or what might go wrong now. But I didn't. I rolled up the rugs and borrowed a kennel from my brother. I bought a fifty-pound bag of organic dog food. I announced that the dog could go anywhere but the couch, a rule upheld so long as I was in the room.

But I'm not always in the room, literally or figuratively, anymore. New jobs and responsibilities and schedules mean the house is empty a couple of days a week, the kids gone to their father's, and often I'm not at my desk when they come home. Frances was still a puppy the first time I saw her recognize the sound of them in the drive. Her head shot up from where she lolled on her bed, and with a yip and a scramble across the hardwoods, she was at the door, practically trembling with excitement, ready to clobber them with her love. They dropped their bags and fell to their knees and let her lick their faces, and since then, this has been their coming home.

Now Frances is a little over a year old. Halfway between puppydom and adulthood, she's full of contradictions. She splays out to take up the whole couch, bed, backseat, and yet she still tries to fit beneath my desk chair when she's frightened. She's got a standing high jump of about three feet but refuses to load up into the car or heft her white ass onto the bed without help. She looks like a seal, a swan, a *Star Wars* character, a bat; we are forever trying to lend her an alter ego, as though she needs to be more than just a dog. Because she's always been more than just a dog for us.

We all share a tiny house downtown. Someday I'll figure out the pound per square foot ratio, but it's high. When a video game or cello practice vibrates through the house, Frances noses into my darkened bedroom, curling up with a dramatic huff on the rug. She'll lick a forgotten coffee cup dry if she finds it on the bedside table. Perhaps the caffeine is a mistake: She looks, often, as if she's been up all night, her eyes red rimmed and doleful. We figure this must be the Saint Bernard.

She's still first to the front door, overly affectionate and protective in turns, sometimes all at once. When she sees my friend Sam, she gets so excited she pees on her own feet. My friend Brian, she will not stop barking, wagging her tail at the same time, her white ruff raised up to show the coppery colored fur beneath it. She won't stop barking even when she's licking his hand.

"She knows how to do her job," Brian says.

He is the kindest man. Frances doesn't know her job from a hole in the ground. She's essentially a teenager, and you don't know anything when you're a teenager, much beyond instinct. Still, you make some pretty important choices on instinct. Who's safe, whom to love. How you take your coffee.

These days, at dinnertime, we scrape the homework to the side of my great-grandmother's dining room table, a gift from my parents in the new homemaking. Meals are simple on school nights, but we eat together. I feel Frances circling our legs (nobody puts Baby in the kennel), and I think how she's kind of like our teenager, the adolescent of this new life we've got here, and it's the only life she's ever known: this tiny house, the three of us, her family.

And I'm ridiculously proud.

My Mother, My Dog

BY DONNA LEVINE

I think my dog is my mother reincarnated. And our relationship has never been better.

I noticed the physical resemblance first. The honey-colored curly hair. The sometimes doleful eyes. The nails. It doesn't hurt that their names–Lulu and Goldie–could each easily be either human or canine.

But it's the emotional side that drove the point home. First, the fear of thunderstorms. Maybe my mother–Goldie–didn't pant and tremble when the rain and lightning started the way Lulu does, but as a child I sensed that this was one of her areas of unease. (I don't, however, remember my mother freaking out around plastic garbage bags. Reincarnation isn't cloning, after all.) I was too young and ill equipped to help my mother. Now I give Lulu melatonin and a little peanut butter when a storm rolls in. It seems to help.

My mother was actually deathly afraid of dogs—until the very end of her life. As far back as I can remember, if a dog was on the loose in the vicinity when she and I arrived at home and were getting out of the car, I had to distract the mongrel while my mother ran inside to safety. We had no garage, so we had to get from the driveway up the walk, up the three porch steps, across the porch, through the screen door, and finally through the inside wooden door. Goldie's fear of these dogs trumped her usual concern for my safety. Daughter be mauled—I'm taking cover! One day Muffin Spencer's Saint Bernard got loose a few streets away. My mother's fear transmigrated into me, and I spent the afternoon feeling as if Godzilla were about to round our corner.

The one time I heard my mother express anything other than terror toward dogs was at the end of her life. She had a brain tumor, and she was living in a care facility called Bishop Gadsden in Charleston, South Carolina. On the phone one day, she told me that some boys had brought in a couple of teacup poodles. "They were so tiny!" she said, her voice sounding the way it did when she described babies, not dogs. "They fit in the palms of the boys' hands!" It was one of the few times she sounded happy in those last months.

Lulu is a cockapoo—half poodle, half cocker spaniel. I had wanted to rescue a dog, as that is clearly a good thing to do, but, having never had a dog growing up (for obvious reasons), I found the shelter overwhelming and the dogs too mysterious: What if this cute pup was abused by a former owner and had neuroses that would come out in all sorts of destructive ways? A friend and devoted cockapoo fan happened to recommend a particular breeder . . . and yes, I played it safe.

I failed to save a shelter dog, and I could not save my mother. She died five months after she was diagnosed. It wasn't until about six years after that that I got Lulu–plenty of time for the reincarnation to take hold.

I was not only unable to save Goldie at the end of her life. When I was twelve, I could not save her from her worries about money after my father left. Then I felt guilty launching off into my independent adulthood. In my late twenties, I told my therapist, whom I had started seeing in part to help me with the panic attacks that gripped me after my first marriage ended and in part so he could help me avoid turning into Goldie, "If only I could get my mother ironed out, I could move on."

I could not save her from her other worries, either. She had joined me on a trip through Europe before I studied at Oxford one summer when I was in college, and she would get so nervous when we were about to leave one place and travel to another that she would have diarrhea the whole night before. Years later, at Bishop Gadsden, embarking on the mother of all journeys, she would beg me and my sisters to bring her more and more Imodium. (Until recently, Lulu threw up every time she rode in the car. I started taking her on small, fun trips–not just to the vet–to provide some positive associations.) Goldie returned to the United States, as planned, before my Oxford program started, so I would be flying home alone. As that day drew near, I had many nightmares about not remembering to pack my stuff, not finding the airport, and on and on.

Even as I inherited whole categories of my mother's fears, I resented her and was unable to reach out to try to comfort her. When she panicked behind the wheel one night because she couldn't find my older sister's new apartment in D.C., I, not yet of driving age myself at that point, sat in the passenger's seat in silence with my arms crossed, lost on my own seemingly unnavigable road. Had we only known that Saint Bernards are famous for rescuing lost travelers . . .

My parents' marriage had been the perfect mismatch: the philanderer and the paranoiac. Unfortunately, Goldie's paranoia did not end with the divorce. She came to distrust me and my three sisters, too, and doctors, and coworkers, and cousins, and sons-in-law, and neighbors.

Lulu goes crazy when the doorbell rings. Only I don't think it's because she's afraid. I think she's eager to see who might be there. A potential friend? Someone to play with? If reincarnation offers an opportunity to improve something from your last life, maybe this is my mother opening up to the world, seizing the moment and a little joy. How delightful to see your eight-pound doggy bounding through life, always happy to start a new day, meet someone new, cuddle on a lap, or doze on the floor by a window in a patch of sunlight.

As geography would have it, I was not by Goldie's side when she died. I was in an airport, anxiously trying to get to her in time.

Since Lulu is my first pet, I have never experienced the loss of a beloved furry friend. Odds are, if luck is on my side, I will have to experience the loss of Lulu. I know I will not feel lucky when that time comes, but maybe I will be able to hold her paw.

I feel lucky now, though, when I can sit quietly with Lulu on my lap or brush her golden curls. I can care for her and sustain her, for now anyway. I hope Lulu's passage through this life is a smooth one. Who knows–maybe one day she will be a fat cat who can sleep through a tornado, and I a tiny poodle who can bring a bit of wonder and relief to the resident in Room 3B at Bishop Gadsden.

Lost and Found

BY LAWRENCE NAUMOFF

It was the mid-1970s, and I didn't know what the hell I was doing, but I had been awarded money by the National Endowment for the Arts, and for a while I didn't have to know what I was doing, because I was, at that time, lucky the way young men can be, and I could be a writer traveling around with an attractive and interesting woman and a good dog and writing a novel.

We were Southerners on our way to Maine by way of Mexico. That requires some explanation. I had just spent a few months pretending to be an expatriate American writer, Graham Greene style, which didn't work out, as I am no Graham Greene.

A friend had bought a farm in Maine, and we were moving there. It had 260 acres with a classic New England house with a steep pitched wood-shingled roof. The house was attached to an L, which itself was attached to the barn, and he had paid $13,000 for all of it. That seemed amazing until we got there and saw how run-down and economically wiped out the whole county was.

Partway up the Pacific coast, we were stopped by what appeared to be two Mexican bandits with rifles. They looked like they were from the film *The Treasure of the Sierra Madre,* from the novel by B. Traven, one of my expat literary heroes, and my dog started to growl. My dog *knew* what he was doing because he had that awesome animal instinct that I later found out, in Maine, is deeper and more inexplicable than I ever thought.

There wasn't another car in sight except for a parked military jeep, and once we were stopped, a clean-cut American guy appeared and checked us and our van for drugs. I had to hold my dog at the side of the road. One of the Rurales pointed his rifle at him.

The other one, who was good looking–imagine Pancho Villa played by Antonio Banderas–was smiling at my wife, and she smiled back, and I thought, it's never going to stop. We had left Mexico because right in the middle of my expatriate writing life in a bungalow on the Pacific Ocean, costing me only seventy-five dollars a week, a young man had his eyes on my wife in a way that worried me, and I decided to put three thousand miles between him and her.

My dog, whose name was Dudley (I've changed his name because he still has relatives down there after that night on the town), knew what was going on with this guy before I did. Dudley was, uncharacteristically, unfriendly to him each time he came around.

So, we were on our way to Maine and I was bitching at my wife for smiling at the Rurales, and the dog was in the front seat, with his head in her lap, and he was, at that time, I now understand, the constant between us. He was, it now seems, so important in our relationship in how purely and equally he loved both of us, and I didn't get it at the time, not at all.

It was a long trip. We stopped in Chapel Hill, North Carolina, and then a few weeks later we arrived at the farm, which is north of Bangor near Dover-Foxcroft. My wife was an irresistible novelty, a friendly, spirited, inviting Southern woman in the midst of the most god-awful, dour, unemotional, shut-down bunch of frowning, suspicious people who were unlike anybody we had ever been around. Even their dogs seemed depressed.

She was like some dazzling, warm angel by comparison, and she made friends with even these people. She thought their stereotypical New England reticence ("Yup . . . yep . . . ayah . . . ay") was cute. To me, they were like characters from the dark side of a novel by Stephen King (who, by the way, was working at the Fogler Library at the University of Maine in Orono and not yet well known). My writing wasn't going well, the money wasn't lasting nearly as long as I thought it would, there was a lot of pressure on me, my artistic license was about to expire, and so we fought a lot.

There were many days when we talked only to the dog, we cuddled only with the dog, we kissed only the dog, we took walks only with the dog. When we were in the same room, he lay down halfway between us, like he'd measured it–he wanted us together.

One morning my wife and Dudley went to town, twelve miles away, down the unimproved dirt road to the "improved" dirt road, and then to the potholed paved road, past worn-out hay fields and cutover woodland and not much else, and finally to the town.

She returned an hour later, and I was writing. She unloaded the groceries and then sometime after that busted into my room and said she let Dudley out when she got to town and then forgot she'd taken him with her and drove home without him.

We headed back to town to try to find him. She was crying, and I was still furious at her and acting like a jerk. We never did find him and drove back home.

There was, at that time, a silence between us and a distance between us that was wider than the three thousand miles I was trying to put between her and "him." It was so wide and so deep, it felt *not* like someone had died, but like someone had been killed and we'd witnessed it, were witnessing it. And it was, of course, us.

Five hours later, Dudley came into view at the far edge of the pasture, directly, as the crow flies, in line with the town, out of the woods, across the thin grass, right on up to the house. He'd walked the whole way. How he did it, where he went, what he went through, I do not know.

We ran to meet him. He kissed us and rubbed against us. We were side by side, he was greeting both of us, he was so happy, my wife was crying again, she was so happy, and my pathetic young-man dumb-shit anger disappeared as if it had never been there.

It's curious, now, thinking about it all these years later, that this dog we picked out of a box of puppies on Franklin Street in Chapel Hill, a whole litter someone had to get rid of, a pup we simply reached down and lifted out of the box and carried home, had become the bridge that diminished the often confusing distance within the relationship, became the constant in our life, an unexpectedly irrefutable connection between us, a brown-and-white short-haired hound of some kind, who came back home, got us together, and stayed the course, and who later, long after he was dead, and without my planning it, showed up in almost all my novels, calmly watching while the humans floundered around in their usual comedy of romance, errors, and excess.

From Cat to Dog

BY BLAIR HOBBS

On the occasion of Twister and Stella's third birthday, I called my then boyfriend, John T. Edge, and invited him over for a Fancy Feast luncheon. He hesitated before saying yes. Eventually, he arrived at my house around noon, and as my two cats lapped their Seafood Medleys from champagne glasses, John T. and I nibbled a few tea sandwiches and strawberries. After lunch, we chatted, and John T. cracked a smile as the sisters twirled in the air, chasing the pink birthday yarn he shook above their heads. Later, I learned that John T. had turned down a more momentous birthday lunch invitation to be with my kitty girls and me. I can't remember which birthday fellow Oxford, Mississippi, writer Barry Hannah was also celebrating, but he was gathering friends at Ajax Diner on the Square, and he wanted John T. lifting a glass at his table.

Eventually, the two sister cats saw us through more courtship, through the early years of marriage, through pregnancy and the birth of our son, Jess. Stella was Jess's nurse. She dutifully ran to him whenever Jess erupted into colicky cries. Twister was his buddy, always at Jess's side for breakfast and bedtime stories.

Months after their fifteenth birthday, Stella fell gravely ill, and while Jess was at school, John T. and I drove her to the veterinary clinic. The doctor told us there was nothing to help her, so we agreed to have her euthanized. Two months later, we took Twister to the same clinic for a checkup, and the doctor told us that she was ravaged with cancer and was very likely in great pain. As with Stella, we were in the room when the lethal cocktail mix was injected, and we tearfully watched her drift off to kitty heaven. That day, during a rare Mississippi blizzard, we buried her in a wooded yard, right next to her sister. It was long after clouds cleared, snow melted, and daffodils pushed through the greening ground that our ten-year-old Jess finally stopped sobbing.

The house felt strangely empty. We missed our four-legged family members–their smile-blinking eyes, their affirming purrs–so when Jess was able to accept another companion, John T. took him to the Humane Society, and they picked out a boy black-and-white tuxedo kitten that we named Eugene Walter, after Alabama's adventurous gourmand. Eugene was far different from the dainty and attentive girls. He quickly grew fat. He hid away in our closet to sleep in balled-up jeans and winter scarves. He preferred to ignore us, but when he did notice us, he often looked as if he were conjuring spells to make our heads explode.

We did fall in love with Eugene, but the love was a distant love, especially for Jess after he discovered that he was, indeed, allergic to this aloof pet. So, talk of a dog began to pepper our dinnertime conversations. "If we had a dog, he'd eat that meatball I dropped," Jess would say as he curled angel hair on his fork. "I'd like a big dog. Maybe a Labradoodle like the Halls'. They have two now, you know," John T. would assert. But I held my feline ground. "Maybe we can get another girl cat," I'd suggest as my gaze floated from my plate to the dead sisters' favorite cuddling chair. And then Jess would tear up, either from residual sadness or from Eugene's dandruff problem.

John T. grew up in the Georgia countryside of Clinton, and he and his parents had their share of requisite dogs. However, they weren't your typical chained-to-a-post-in-the-dirt-ground kinds of mutts. They owned and pampered two Old English sheepdogs, one named Trilby and the other Elton. And for several years, John T.'s dog companion was a little white Yorkiepoo who shows up as a blur in many of the family photographs. Her name was Lilliput, and to John T.'s horror, she wore a diaper.

With Elton, Trilby, and Lilliput informing John T.'s breed preferences, it was no wonder that he was especially fond of big, manly dogs. We'd walk home from the bar, and John T. would spot a jogger, a big boxer trotting by his side. And of the breed he'd declare, "I like that dog. He's a *real* dog." Or of our neighbor's yellow Lab who routinely escaped his electric fence and roamed our yard, John T. would muse from the living room's window, "Biscuit's a fine dog."

I was losing ground with my argument for another cat. I was painfully aware that Jess was growing up quickly and that if I didn't act on their wishes, he would be deprived of that all-boys-must-have-a-dog notion, and it was going to be my fault. So, late at night, after Jess and John T. had gone to sleep, I scanned the

Lafayette County Humane Society's Facebook page, where there was a listing of pets in need of "fur-ever homes." The listings were paired with photographs, and they were all impressive.

A few nights into my dog stalking, I fell upon a photograph of a dog described as an Italian greyhound/rat terrier mix. Her eyes were chocolate brown, her coat black with a ghost-shaped swirl of white on her chest. As I lay in bed that night, I couldn't clear her face from my mind. Her ears draped flat to her head; her eyes were sweetly sad. Her countenance vaguely reminded me of a Mother Teresa picture I'd seen in Sunday's *Parade* magazine when I was a kid.

I showed Jess and John T. her photograph, and although they thought she was cute, she didn't quite fit the bill, especially for John T., who was hell-bent on a manly man's dog and not another Lilliput. "We'll keep looking," he assured.

In a state of petulance, I posted the homeless dog's photograph on my Facebook page and launched what turned into a campaign to change John T.'s mind. Soon, the comments morphed from "Oh, that's too bad" into "How can John T. be such a heartless hater?" and within a day folks began to barrage him with phone calls and e-mails, trying to convince him to change his mind. To me, these well-intentioned folks would instruct, "Just go and *get* her." Alas, the campaign backfired. John T. was—as he should have been—quite angry with me.

Then something happened. Maybe John T. stole a few glances at the Humane Society's online photograph, or maybe my cruel campaign actually worked. I don't know what exactly changed John T.'s heart, but two days later, he surprised Jess and me by suggesting that we *all* visit that shelter dog I'd "gone on and on about," that we would make this decision "as a family."

It was chilly as we walked from the parking lot to the Humane Society's kennels. As we approached the outdoor enclosure, our dog of interest led the charge to greet us at the chain-link fence. We were shocked by how tiny she was, and her gill-like ribs were startlingly prominent. She leaped up, crossed her paws through the fence, and with her fruit-bat tongue she made her best effort at bathing our fingers with a million kisses.

The shelter worker greeted us and led us inside the gate and into the building. We tried to get a good look at the "Italian greyhound/rat terrier mix," but she moved too quickly and bounced like a small black tumbleweed across the cement floor. Even as Jess tossed the frantic dog a tennis ball, John T.'s eyes seemed to scan the cinder-block building, its cages, to light upon his longed-for sturdy dog. And I have to admit that even I was shocked by how spastic this dog was. As she jumped for the tossed tennis ball, all fours desperately churned the air, her body twisted crazily, and her floppy ears flew straight out, making her head look like a big wing nut. When the tennis ball bounced behind a carton of paper towels and was no longer in muzzle's reach, the scrappy dog sniffed the cold floor and looked up at us as if we had stolen her only friend. John T. scooped her up into his arms and stroked her trembling head, and I felt something similar to buyer's remorse even though we hadn't yet bought anything. As we walked across the asphalt parking lot to the car, I echoed John T.'s words: "We'll keep looking."

The following Saturday morning, I graded essays and Jess played Xbox games. John T. was in a good mood–whistling a no-tune song and frying up extra bacon. The morning sunlight poured through the kitchen windows and washed away all of the week's familial tension. After a lazy brunch, Jess and I resumed our tasks, and John T. left for his usual errand running. He was gone for a good hour or two, and instead of returning with a farmers' market bounty, he came home with our dog, the little dog for which I had so disrupted our home life. I think Jess froze into a mild state of shock when John T. took the game controller out of his hands and plopped the dog onto his lap.

We named her Lurleen Wallace after another notable Alabamian, the former first lady and governor of Alabama. Although Lurleen was no bigger than our housecat, she proved to be all dog. Within days, she chewed through rugs, the chesterfield sofa, two club chairs, and a fancy silk pillow, and she shredded an already pitiful bromeliad and hid it under our bed. She volunteered as Eugene's personal trainer, chasing him and mercilessly pulling him around the house by his ears. She barked at the mailman. She ate garbage. She was wonderful. John T. and I were both delighted by our bonkers girl.

Initially, Jess was shy around the little dog. I suspect that he was worried for his heart, knowing that if he did fall in love with her, and if we leave this earth in the order in which we're supposed to line up, he'd not only have to bear away another cat, he'd also have to say goodbye to his first dog. But Lurleen's persistent need for being needed proved irresistible for even our guarded boy. At breakfast, Lurleen learned that if she sat still at Jess's feet, he'd eventually slide her a bite of sausage. In the afternoon, when Jess came home from school, she exuberantly pranced around him like a junior-high cheerleader.

One Saturday evening, Jess stepped into the utility closet and fished out an old can of tennis balls, popped the lid, and directed

Lurleen, John T., and me into the backyard. John T. and I sipped wine as Jess tossed the neon ball across the grass. Lurleen followed and ran more quickly than we could ever imagine, churning the yard's distance with astonishing greyhound speed. When she caught the ball, she returned it to Jess for another throw. Over and over, they played fetch until Lurleen happily surrendered beneath a starlit hydrangea. We all retreated indoors, and when Jess shuffled to bed, his panting dog followed. Before turning in, John T. and I peered into Jess's night-lit room and couldn't help but smile as we noticed Lurleen–instantly loyal and ever sweet–curled into the crook of our boy's tired arm.

Canine Couple

BY BUCKY McMAHON

Are they brothers?" people on my suburban street sometimes ask. Actually, no, my mutts, my brindled pit bull mixes, are husband and wife, so to speak, and possibly cousins, given their partial resemblance and same-pound provenance. Tigre is the wriggling one beaming love like a born fool. Almost as striped as a tiger, he can resemble either a sleek sea lion with bathetic don't-club-me eyes, or a bristling hyena–the two sides of his personality (fortunately, only cats and varmints see Mr. Hyde). Sylvie is a stocky black dog with copper and silver brindle, growling a little, but hackles going down, tail beginning to wag. Initially, she's suspicious, but if you pet her, she will begin to lick the air as if the whole world has turned savory, and maneuver her backside to be whacked, which is her perverse bliss. Alas for her, no street-side encounter has ever advanced so far.

Alas for me, I'm never invited to expand upon their unique and absurd histories, which would be my own version of licking the air. Instead, I inquire about *their* dog or dogs, find some quality to praise, and we move on, or they do. Tigre must first perform the Tigre dance, some ritual kicking up of turf and pissing in the four directions. I feel like a Christian missionary when I have to break up these rites to protect someone's begonias. Meanwhile Sylvie begins a thorough crime scene investigation of curb and tarmac, sniffing and tasting, and sniffing and thinking, brooding like a canine Sylvia Plath composing a poem in her head. She cannot easily be rushed to a conclusion.

These idiosyncrasies can make the walk a chore. I try to be patient and amused. They are country dogs adapting to town living. Tigre in particular has come from prehistory to the modern age—from nature red in tooth and claw, to parks and poop bags—in a twenty-minute car ride. He's a Huck who's been civilized. But unlike Huck, he's got gray on his muzzle. Sylvie too. They have outlived their country idyll, and outlived their pack, too. They are dogs with biographies.

Once upon a time the pack was nearly a dozen. But that was in the Caribbean, and as I pointed out repeatedly to those who complained, only three of them were our dogs, the Irish setter and the bearded collie and the Rottweiler puppy I found abandoned by the road as I was walking home from surfing. Heather and I were practically newlyweds, rich in many things, including dogs. A few years later, in the heartbreaking way of the world, we inherited a fourth dog, Heather's mother's standard poodle, Pepita.

By then we'd left the islands and settled with our pack of four in relative bucolic bliss on fifteen rural acres in North Florida. For Tigre and Sylvie—not yet born—those four dogs would become the

Greatest Generation, pillars of pedigreed dogdom with none of their muttly flaws. We called them the Cartwrights, for so they would comically appear, coming shoulder to shoulder through the tall grass of the back meadow. The bearded collie, Rebel, was the patriarchal silver-haired Ben; the slim black standard poodle, Pepita, a dead ringer for Adam; our Rottweiler runt from Puerto Rico, Faro, was a jaunty Little Joe; and Penny, our food-obsessed Irish setter, was Hoss, of course.

There was never a problem with walking the Cartwrights without leashes through the woods or along our gravel road. All day they lay about in sentinel dugouts under the shrubs. But dawn and dusk, they rose to the occasion of a good long trail walk. And they followed our lead as good dogs will. They came when called. They weren't trained. They were just normal, level-headed dogs enjoying the freedom of country living, and so time wagged on for many a year, letting sleeping dogs lie.

Of course there would be no *Bonanza* without episodes, and no episodes without outsiders riding in to the Ponderosa. The first guest star was Bounder, so named for his vertical leap in the animal shelter pen. He leaped and leaped again, and if you passed on would leap some more, maintaining hopeful eye contact over the six-foot wall. I was impressed by the energy and character of the year-old brindled Staffordshire terrier with the crudely cropped ears, and thought it would be a privilege to observe that energy unbound.

It had been raining hard for days, the creek had jumped its banks, and the whole back parcel was flooded the day Bounder came home to us. We set out for the morning walk in rubber boots with the Cartwrights and the newcomer leading the way. A game of chase ensued, a swamp romp the likes of which we'd never seen. Back and forth Boundy Boy flew, raising a rooster-tail wake. Who could catch him? Nobody! No way! He surged and plunged, sprinting through the shallows, paws beating a tattoo like a crazy bongo drum. We laughed out loud. This was all right, eh, Boundy? A bit better than the pound?

But Bounder was not an easy dog, and not a lucky one. Six months later he was killed by a monster diamondback rattler. I was devastated, and still reconcile myself to the loss by remembering that if Bounder hadn't died, Tigre would have. Friends at the animal shelter who knew I'd lost a brindled bulldog called to say there was a puppy, yet another of those "pit bull mixes," whose time was nearly up.

At three months Tigre was a lithe, low-slung creature like a badger–and he badgered the Cartwrights to play. He adopted Penny, the wide-body Irish setter, to be his new mother, and joined her pond-side frog hunts. Penny's chances of actually catching a frog dropped from a hopeful zero down into the negative integers. As serious as she was about stalking those frogs, she grew frantic seeing her career sidelined by a rookie. In truth, none of the old guard were all that crazy about Tigre. Nor did he stay out of trouble. Within the pack he fought once, he fought twice: He became known as "the dog with two strikes." As his attorney, I cited his shelter background and argued that he was a good dog at heart. I'd seen him break bad with a neighbor dog, a ruffian named Cracker–who started it–and I swear Tigre cried as he fought. From very complex emotions, I would say. But on the whole, and as he began to grow out of puppyhood, it was clear that he'd been poorly received, tolerated at best by the Cartwrights, which was sad because he wore his heart on his sleeve. I began to negotiate for one more dog, a therapeutic young companion.

There's never a shortage of "pit bull mixes" at the pound, even brindled ones. If you want to choose the least likely to be adopted, take home a black one, like our Sylvie. She was about Tigre's age, about his size. But whereas Tigre was genetically a hunter, and a terrier–with that compulsion to dig into the *terre* to root out burrowing critters–Sylvie was a herder, with some cattle dog in her, some chow, maybe, and a curiously somber and miniaturized face, as if there were such a thing as a toy Lab. That face has never grown, though the dog behind it has expanded considerably.

The early years of their marriage were difficult ones. Tigre maintained the provoking tendency to move freely, chasing scents as they wafted by, as if he were an independent agent; while Sylvie saw at once that he needed to be monitored, his peregrinations checked, and, if possible, curtailed entirely. For the nuptial couple, walks became a series of ambushes, Tigre headed off at every pass, both eventually frozen in a standoff, trembling with anticipation. Make your move, hombre! With a high-pitched squeak, half exasperation and half yippee ki yay, Tigre would juke left, break right, sprinting for home with his girl dog nipping at his legs. Ah love. And yet for all their perambulatory dysfunction, they would often lie together in the sun, curled like a brindled ideogram, a canine yin-yang.

Time passed, and this being a dog story, dogs died. The limestone markers in the shady part of the vegetable garden multiplied. Tigre himself nearly died of a horrendous water moccasin bite to the tongue. It was a near thing, requiring an emergency tracheotomy. I prayed all night as I hadn't prayed before and haven't since. When Tigre pulled through, recovering with just a conversation-piece notch in his tongue, he became the true dog of my heart, my all-time fave—by a nose.

During the years in the country with just Tigre and Sylvie, I would say that they grew apart. They had their moments of closeness, but essential differences in character urged them to separate pursuits. For some time Sylvie undertook to herd our vehicles, which could be slowed if chased, stopped if headed off at the windy end of the drive. From this time of delayed departures, she acquired another nickname, "the Worm." She was hard to catch, and when caught could worm out of her collar, what with that tiny head and all. One rainy morning, amid some passionate cursing, a sizable branch came flying at her. Sylvie had a rethink then, tapered off and then dropped the practice of herding our vehicles. By steady increments she grew increasingly companionable to me, shadowing me by day, sleeping at my feet as I wrote. I can say that she passed her final years in the country mostly in reflection, pondering what it is to be a dog, and just to be.

Meanwhile Tigre took up ratting at the new next-door neighbors' barn-like garage. If they would just clean it up, I would argue as his attorney, maybe he wouldn't have to spend all day and all night patrolling for vermin. In addition, the new neighbors had an indoor dog, a very attractive fellow, hunchable, in a word. So Tigre generally figured he'd wander on over for a little ratting, a bit of hunching. And furthermore, these neighbors would leave, and come back. Fancy that. Wouldn't it be reasonable to think that if they left, they would not come back? And if they did, a certain amount of barking would be justified? As for those incidents at 3:00 a.m., the lusty hue and cry, the blood lust and shell crunch of armadillo, I'd rather not say more. That was the Tigre story. If there was trouble a dog could get into, he got into it. All this made him very happy.

The Cartwrights were a halcyon memory, as were the twice daily woodland walks. In all their years, our pedigreed dogs never, ever, considered not following us into the woods to walk the trails. Unthinkable! And yet after only a few thousand circuits, the mutts gave it up. Dropped it flat. What was back there anyway? Not rats, Tigre knew, not bromance. Maybe you guys, Sylvie acknowledged, but she knew us. We'd be right back. Au revoir. Enjoy. It was the damnedest thing. Around the time they lost interest in following us on formal walks, they became known, theoretically, potentially, as "our last dogs. Maybe. For a while."

But truth be told, we'd lost enthusiasm for country living almost simultaneously with "the last two dogs." It was time for a change and we moved to town. We worried about how it would affect the dogs, the irrepressible Tigre and the pensive Sylvie. Not to worry; their roots were urban all along. The mighty Thunder God looms and dooms country and city dog alike, but otherwise the occasional siren and hiss of traffic are no bar to Sylvie's deep meditations. For Tigre, ever at the door, notched tongue panting for the next excursion, town is a paradise of rat-able, cat-able, hunch-able opportunity. All foiled so far, but next time!

So off we go, dawn and dusk, as in olden times, the days of the Cartwrights, only as townies now, with leashes. Tigre's at the door. Sylvie's coming too. Sylvie! You gotta go. (She's gotta go.) And we'll see a great many cats and a great many dogs, and people, and people with dogs. Tigre will cry—he cries with joy—and Sylvie growl her low growl of suspicion. But as we approach those people we decide are approachable, the pair will synchronize their greeting wags, because they're both nice dogs, really. "They're beautiful," people say. "Are they brothers?"

CHAPTER 5

Life Lessons

Swim Team

BY DOMINIQUE BROWNING

That dogs are emotional creatures—that they keen with sadness, leap with joy, wiggle in friendship and waggle in play, that they endure heartbreak and separation—is a thing well known to all who count these fascinating creatures among their best friends. But that dogs suffer neuroses? That they can be crippled by anxieties buried in their tribal past? That fears as deeply buried as last year's bones can ooze up out of the primordial pink of their brains, and keep a dog from being all the dog that she can be? Or that their genetic pools might get as twisted and polluted as ours do? I had no idea, until I met Ozzie.

There was something off about Ozzie from the moment she entered the lives of her owners—dear friends of mine, so I was able to observe her case at a close distance over many years. She was adorable; what puppy is not? A chocolate Lab, full of bounce and energy, Ozzie never met a shoe she did not chew. But what was cute in a baby was less so in an adolescent, and intolerable in a young adult. Dogs, as you know, have a way of bounding through these stages at warp speed.

Ozzie seemed untrainable. It wasn't because of poor parental skills, lack of discipline, hazy directives, or old-fashioned spoiling, all of which were, to my critical eye, on full display. Those sweet, indulgent friends of mine did finally send her off to boot camp so that she might have a chance to pull herself together. And it worked, sort of. Still, Ozzie could not stop chewing. Chewing herself, that is.

And she was chronically depressed, putting on weight, listless, turning her back on all the doggyness the world had to offer. Finally, she was diagnosed with obsessive-compulsive disorder. Before too long Ozzie was chowing down Prozac, or Wellbutrin, or some such chemical cocktail, so that she might have a chance at a full and happy life.

As I mentioned, Ozzie was a Labrador retriever. We were living on the coast of Rhode Island, and we went to the beach every day to walk and play. When Ozzie was a puppy, I looked forward to doing that thing everyone does with Labs—throwing balls and hanks of driftwood into the ocean, watching as their Labs hurtle across the sand, dash out into the foaming surf, and *retrieve* the things, jaws clamped down purposefully, commandingly, and, then, battling the break of waves, riding the swells, return joyfully triumphant to drop the thing at their owners' feet, gaze soulfully up, brimming with quiet dignity, tail wagging proudly to do it, please, just one more time.

But Ozzie was afraid of the water. Very, very afraid of the water. She wouldn't go near it. She would bark at it, from a safe distance.

She would paste herself to our legs when we walked the shoreline, making sure she stayed on the far side of the surf. She made a half-hearted attempt to chase a gull or two, but they seemed to know instinctively (as if you couldn't tell by looking) that Ozzie was no threat. It was pathetic.

This was when I began to suspect that something was really wrong. But as I said, she wasn't mine; and as anyone knows who has reared her young, you don't get a lot of points for being a backseat parent. Besides, she was already on drugs to control the OCD thing. We weren't expecting to take her, or ourselves, for that matter, hunting for ducks, so what did it matter if she wouldn't and couldn't swim?

It mattered to me. There is almost nothing I like more than the heavy, satin feel of cool ocean water against my bare skin. All winter, I remember the way that water caresses me, and holds me and lifts me in its swells and then parts for me as I glide through it. I pine for the ocean. I long for spring, when I can go in up to my knees. I swim every day as late as I can into autumn.

I envied Ozzie's thick, oily coat, which seemed impervious to frigid temperatures; I wanted those large, powerful neck and shoulder muscles. Ozzie had all the right equipment. But she was wasting it. And besides, I don't believe that drugs are the answer to everything. I believe in talk therapy. I knew that quite often, depression is born of powerful inner conflicts that cannot be expressed, much less resolved. Ozzie seemed to like talking, though it didn't help her swimming problem.

It was vexing, to see Ozzie suffer as she hit middle age; vexing to watch her backslide and attack her paws, vexing to trip over her at the beach whenever the water dribbled too close. I am not the sort of person who can turn her back on other people's problems. In fact, I like nothing better than a mess to clean up.

One day, I had an idea. I decided that somewhere deep down inside, Ozzie knew that she was a swimmer. She knew she was supposed to frolic in the surf; she knew she was meant to pull through cold water. Ozzie knew all this, in her doggy, Labby soul. She was stuck, and becoming ever more entrenched in denying her true, retrieving, water-loving self. And what is neurosis if not the unhealthy repetition of actions that make one unable to be one's true self? Fear had won the day–and this had embarrassed and humiliated and shamed Ozzie. She could not see a way out.

But I could. I would teach Ozzie to swim.

We went to the beach. It was a still, cool day. The ocean was calm and the light on the water dazzling. There was only a gentle breeze, and the surf came in with hardly a ruffle. Ozzie and I wandered away from her family so that we could be alone. She trusted me; after all, she had watched, fearfully barking and whining, for years, as I had gone into the water, paddled around–and returned safely!

When our privacy was ensured, I sat down next to Ozzie in the warm sand. I began to talk quietly to her about the water, which cloaked our conversation in the classic white noise of the psychiatrist's office. I told her how much I loved the water, and how good it felt. As I talked, I moved slowly, gently, closer to the tideline. Ozzie huddled next to me. A bit closer, and the water was lapping at my haunches. Ozzie's attention seemed riveted on my face, as if she knew that something important was about to happen.

I held her paw in my hand, as we sat, and as the waves came up, she flinched a bit–I could feel a shudder pass through her–but then she settled down. The waves kept coming back, as is the way of waves, and the tide began to rise, as is the way of tides. But still we sat. Before too long, we were sitting in a few inches of water, just holding paws. I coaxed Ozzie into a lying-down position, and I got onto my own tummy. I had actually never simply lay still and let the water ripple around me, and it felt surprisingly lovely; I was clearly having personal breakthroughs of my own.

Soon the water was high enough to give me some buoyancy. I pinned my elbows into the sand, and faced the waves, tadpole style. Ozzie scrambled to her feet, but she stayed next to me. I moved a bit deeper into the water, and Ozzie began wading. I laced my arm around her leg and up under her neck. She knew I was right there with her. She was knee deep, then gut deep, determined to stay on her feet. I let a wave push me up against her, and gave her a little bump, just the tiniest shove, and Ozzie was off her feet–just for a few seconds, until the wave retreated. This happened eight or nine or ten times. Ozzie let herself become buoyant, and then she settled her paws back into the sand.

And still we talked. About the nature of life. About how waves came in, and went out, because that was what waves had to do, and about how people, and dogs, and ducks went into the water, and came out again, because that was what dogs and people and ducks do. (I don't know why I brought the ducks into the picture; perhaps it would trigger some atavistic response.)

By then we were in pretty deep, as far as Ozzie was concerned. I took her paw again, and on the next wave, the next moment of buoyancy, I crooked her leg, and showed her how to paddle. She backed off, so I told her I would do it, a little dog paddle just the way people learned to swim, and just the way other dogs did it. Then I hunched down next to her, and took both her front paws, and paddled them through the water, to give her the hang of it. I could feel her heart pounding against mine.

Ozzie was getting excited, I could tell. A spark lit in her eyes. She looked at me, and I smiled. She smiled back. We started moving along, keeping shallow, in a combination of crawling, walking, and paddling. Before too long, we were in far enough that neither of us could crawl or walk. Before too long, we were swimming. And I was cheering. "Wonderful, Ozzie. Good dog, Ozzie. You are doing it, Ozzie. That's fantastic. Beautiful, good girl, Ozzie. What a great dog you are" and all the other things, exaggerated with love and pride, grown-ups say to small ones who are conquering new frontiers. I stopped holding her paws to help her; she moved rapidly out of my training arms.

Ozzie was swimming.

On the shore, her family had gathered, and they too were cheering excitedly. Ozzie glanced around, a new fortitude in her shoulders, but she was concentrating too hard on her new skills to pay much attention to their shenanigans. I stayed right next to her, dog-paddling too, bobbling against her side, talking, cheering, being present. She yipped back from time to time. We talked about what it meant to give in, to relax, not to fight. About what it meant to let go of fear.

And that was how Ozzie learned to swim. And as far as I was concerned, that was how Ozzie finally became . . . more of the dog she was meant to be. She would never be a duck hunter's trusty companion. But Ozzie had retrieved just a little bit more of her soul.

Puppy Dreams

BY PAT JORDAN

Susan is seventy-two, I'm seventy. The Lady Sashay will be our last puppy. Over twenty-three years, we've buried six dogs and have four left: Sweetness, fourteen, with bad knees; Matthew, nine, a bastard out of Carolina we rescued; Nicholas, three; and his brother, Precious, two. Matthew is a retriever-spaniel mix, our designer dog, a Retraniel. All of our other dogs have been Shiba Inus, Japanese hunting dogs. For centuries, the Japanese used Shibas like pit bulls, to hunt bears and wild boar in packs, and in dogfighting. A dogfight between Shibas never ends in whimpering capitulation; it ends in death. Shibas are tough, smart, obstinate, primitive dogs, more like small wolves. Mark Doerr, a noted dog behaviorist, told us that Shibas "are the dogs impossible to train." When you feed any other dog, he will look at you as if to say, "Oh, you're feeding me! You must be God!" When you feed a

Shiba, he looks at you as if to say, without exclamation points, "Oh, you're feeding me. I must be God."

Like most hunting dogs, Shibas are aggressive toward animals but see humans as their allies. Shibas love people, as long as they behave. If not, they turn their backs on them. It's a humbling experience, the first time you're condescended to by a thirty-five-pound red-and-white dog. (I am remiss here in not writing about Matthew. But he does not have the depth of personality, the range, the bite, you might say, of our Shibas. Our Shibas expect us to worship them. Matthew wants only to worship us. A simple creature close to God.)

Male Shibas are playful and loyal. Female Shibas are aloof, like cats. The Japanese call Shibas "the cat dog." Our first breeder, Linda, told us, "They don't call Shiba females bitches for nothing. In a pack of males, a lone Shiba female is the alpha. Shiba males always defer to the bitch." Our first female of two, Kiri, was charming with people, like Perle Mesta at a soiree, but bossy with her older brother Hoshi. Our second, Stella, was yappy, foul-tempered, and belligerent, even as a puppy the day we picked her up in a cargo hangar at Delta DASH. A baggage handler carried the doggy crate toward us; the crate was rattling and the creature inside growling and snapping. He held the crate stiff-armed away from his body and said, "Whatcha got in here, a wolverine?"

Stella never stopped yapping, growling, and snapping for sixteen years. She was always pissed off because a human or a dog crossed her sight line, her food wasn't delivered on time, or one of her brothers had the temerity to drink from the water while she was. She would snap at him. He would back off, sit, and wait. I think Stella's pisstivity was a defense mechanism to hide her fears and insecurities (you tend to do this with Shibas, psychoanalyze their neuroses), because she never bit a dog or a person. She was most afraid the day she died. Susan sat on the floor, Stella climbed into her lap, the first time ever, and stared up at Susan, as if to say, "I feel strange. What's happening to me? Make it go away." But Susan couldn't.

After Stella died, we swore off Shiba females and settled in with our four boys to a Stella-free life. But last summer, with our boys grown indolent, we decided to jump-start them out of their pastoral torpor with a bitch. So I called Fred, our new breeder in South Carolina, where we now live, and asked him to save us a bitch from his next litter. "Fine," he said, a man of few words. He didn't question our getting a puppy as our friends did: "At *your* age?" We'd already thought about that. If we didn't live to bury all our dogs, we left those remaining in our will to Peter, our dearest friend in Miami. Peter's a dog lover like all our friends. Susan and I have always loved dogs more than people—our curse, or blessing.

When Fred e-mailed that one of his champion Shibas had had a girl, we e-mailed him that I'd drive down from Abbeville to pick up "the Lady Sashay" in six weeks. We liked to get our Shibas at that age because they were still unformed puppies, a little wild, but malleable, their personality not yet hardened in stone. At six weeks, they were more easily enfolded into the pack because they only wanted to do what the pack did. At eight weeks, Shibas have already developed their arrogant Shiba attitude that could cause trouble in a pack of older Shibas set in *their* ways. Shibas can be aggressive toward other dogs, even those in their pack. We'd broken up our share of dogfights over the years, over dog food, a possum one of them killed and the others wanted, a perceived doggy slight, an insult, whatever. We got tuned to the telltale sound of low, throaty, threatening growls followed by a slashing of teeth and blood.

Every Monday, Fred sent a photo of the Lady Sashay. (It's always "the Lady Sashay," never "Lady Sashay" or "Sashay.") At first she looked like a gerbil, then at four weeks a Shiba puppy: pointy nose, pricked triangular ears, tail curled over her back, a dusty reddish color, white muzzle, white chest, white paws. At eight weeks, she'd turn fiery reddish brown.

We showed our friends the photos. They asked, why the Lady Sashay? We said, "A fitting name for a Southern girl, no?" We talked about her as if we already knew her. She was a Shiba; we already did know her, up to a point. Each week we invested more of our emotions into her, knowing she would never betray that investment, as people had. We saw her in our mind's eye, her haughty little ass-shaking walk; "I'm me, and you're not!" hence Sashay. We anthropomorphized her as we had all our Shibas, a typical failing among Shiba owners. The Lady Sashay in her white stiletto pumps, flouncy miniskirt, little tube top, big white-framed sunglasses, her ponytail swishing as she licked a lollipop. We looked down at our sleeping boys and thought, "Life as you know it, boys, will soon be gone forever."

The day before my three-hour trip south, I packed Susan's knit purse with towels we'd rubbed over our boys' fur so their scent would be the first thing TLS smelled in the car. I planned to put the purse straps around my neck so it would rest against my chest, with TLS inside and my hands free to drive. I put the purse, puppy toys, pink rhinestone collar, blue leash, ziplock bag of puppy food, and metal water dish with a bottle of Evian in the car. I e-mailed Fred that I'd arrive the next day before noon, then rechecked my route, to Paxville and Fred's kennels, Frerose Shiba Inu. Fred was the most famous Shiba breeder in the world, outside Japan.

Finally, I went to bed. We watched movies for a while, but I couldn't concentrate, so I went outside to the porch and began to pace to calm myself. When I exhausted myself, I went back to bed. The next morning I saw Fred had e-mailed me. "Hi, Pat, we're sorry to tell you but you can't take Sashay home tomorrow. Her eyes are only half open, not uncommon in oriental breeds. Her head has to fill out for her eyelids to open completely."

A week passed. Another. Susan and I began to imagine things: The Lady Sashay had an eye infection, her vision was bad, her head had not developed. When I e-mailed Fred about these fears, he wrote back, "It's just her eyelids, she'll be fine soon." But soon was not soon enough for us. When TLS was eight weeks old, I e-mailed Fred, "I'm sorry, Fred, but we're gonna have to pass on this puppy. It's important to get one at six weeks. We'll wait for a female from your next litter." Fred understood.

We took down the Lady Sashay's photos from the refrigerator. Susan put them in her desk drawer; she didn't have the heart to throw them out. I put her leash, collar, bowl, toys, food, the purse, into a plastic tub in the attic.

We tried to rationalize our guilt for rejecting her. "At nine weeks she could have disrupted the pack," I said. Susan nodded, fighting back tears. We tried not to talk about her. When we did, we called her "the puppy." We struggled with something profoundly upsetting we could not define. It was not rational. We had read into that puppy what we read into all our dogs. They were objects that fulfilled our need to nurture, love, trust, protect, make happy. Now the Lady Sashay had betrayed us. Irrationally, insanely, I was furious with that poor, innocent puppy with the droopy eyelids. I had told Fred we didn't care about the eyelids, we wanted her at six weeks, we'd care for her, we'd love her—but he said no.

Weeks passed. We no longer talked about the puppy we'd rejected or the one who would replace her. Then one day in October Fred e-mailed that he had a female puppy for us we could pick up, at six weeks, in late November, "if everything's all right."

"A Christmas puppy!" Susan said with feigned brightness. "We could call her Noelle." But we didn't talk about her much. When we did, we called her "the puppy." Then one day, I said, "I was thinking, if Fred hasn't been able to place the Lady Sashay with anyone, maybe I should take her home with the other puppy, too."

Susan said, "Absolutely."

Last Rites

BY MARY LOU BENDRICK

When I tell people that I had my dog's body frozen, I have to assure them that it was not a cryogenics experiment.

I try to explain. "Because of the funeral."

That's usually when eyebrows rise, especially if the listener is a Yankee.

My Yankee friends prefer a quick, clean cremation for their pets. It's not a matter of being thrifty–my friends spend more money on organic kibble than they do on shoes. I believe it's because those from Up North have never had a good association with a funeral. I can personally attest to this: I am a Yankee Catholic girl, so my funeral memories involve somber, pinched events with bad food.

I took an interest in Southern funerals in 2005 after reading *Being Dead Is No Excuse: The Official Southern Ladies Guide to Hosting the Perfect Funeral* by Gayden Metcalfe and Charlotte

Hays. I was sick with jealousy to learn that a proper send-off could, and even should, include good food and a restorative cocktail. I am not proud to admit this, but at my grandfather's funeral I was angry because the cake was bad. I can attribute my attitude only in part to my being a self-centered teenager; the cake was inexcusably dry.

In 2006, my dog, Simon, became seriously and mysteriously sick. During that time I visited friends in Mississippi. There I met Barbara Davis, a woman who told me about a funeral she'd held for her dog. When Moonpie, her springer spaniel, was taken too soon by an adrenal tumor, she didn't just pack the body off to an incinerator. Barb commissioned a custom-made heart-pine casket so heavy that it required four human pallbearers. There were so many funeral details to tend to that Barb forgot to close the heavy pine lid, resulting in an impromptu open-casket affair. It all worked out well enough: Mourners placed toys, flowers, and ribbon-wrapped hot dogs next to Moonpie, who rested on a large pile of Dollar Store pillows (the casket was so deep that the diminutive deceased initially looked as if he were lying at the bottom of a well). A bona fide Episcopal minister delivered the service, during which a bereaved golden retriever keened. Among other more serendipitous details, a black GMC Jimmy served as a hearse to transport the backbreaking casket from the front door to the backyard, where a tented grave had been dug with a backhoe. The residents of Brookhaven, Mississippi, did not look askance at a canine funeral. In fact, some were offended that they weren't invited. Barb had to explain the snub: "You didn't even like Moonpie!"

Moonpie's funeral sealed the deal for me: Simon had lived

large and deserved a grand exit. In his heyday my German wire-haired pointer–black Lab cross was handsomer than George Clooney, with a distinguished beard. His physical beauty, though, was frequently marred by the presence of a head cone, and his thick vet file was a testament to a full life. In thirteen years he racked up thousands of dollars in medical bills for mishaps such as deeply embedded porcupine quills; a near-fatal beaver bite; a broken tooth (a lower incisor caught on a tug toy); seizures from consuming an entire deer gut pile; and cancer that we caught just in time. He'd also been in an avalanche and hiked to the top of a fourteen-thousand-foot peak. He once saved me from a charging horse—a gelding whose late-life castration left him without testicles but full of stallion attitude. "Rusty" happened to be in a large pasture when I entered with a mare. The hormonally deranged beast pinned his ears and ran toward me, intent on harm. Simon intervened, hackles up and teeth flashing. Rusty veered off.

The flip side of this heroism was embodied in the Post-it note stuck to Simon's vet file that read "No Cotton Balls." A smart animal, this dog had figured out that cotton balls presaged a needle. The mere sight of white fluff sent him into a wild-eyed panic that required the assistance of the entire veterinary staff and a chicken-flavored sedative or three.

Ironically, what got Simon in the end turned out to be something very small: a tick. Lyme disease laid waste to his aging body with a cruel succession: a limp, followed by arthritis in his spine and hips, then seizures unrelated to dining on deer guts. Despite a few trips to a university vet hospital (the vet file grew in proportion to my shrinking bank balance), no one could explain why one eye was now permanently rolled back. His long gray beard supported the waxing zombie look. That he walked like a drunk did not stop him from attempting to ambush squirrels from beneath the porch, where he kept vigil with that one good eye. The neighborhood kids called him the General. Even enfeebled, he was the stuff of legend.

After much anguish, we sent him to his great reward on a hot July afternoon in the comfort of our living room. We distracted him from the vet's lethal needle with ground beef, which we later referred to as the "Judas hamburger." His long-standing request was honored if entirely unnecessary: no cotton balls.

Amid our sharp, raw grief came the logistical problem: We wanted to bury him in the pet cemetery on a family property in Vermont, several hours from our house in Massachusetts. But it was a Tuesday and my husband couldn't take the time off from work until the weekend.

It was our vet who came up with a solution (you become close with your vet when you pay off his medical school loans).

"I'll just freeze him for you," he said nonchalantly. "I do it all the time."

Evidently, vets routinely store dearly departed pets in big freezers until the crematorium truck arrives.

So my late dog's body was molded into the shape of an eternal doughnut and placed in a giant freezer. A cousin kindly offered to drive the body to Vermont so that our children would not have to ride with the frozen corpse in the back of our station wagon. (I'd wanted to transport the body myself with the intention of speeding the whole way, hoping to be pulled over. Imagine the look on the police officer's face when I explained the urgency.)

We buried the General without incident in a deep grave on a sweltering day, an ache rising in my chest like the thunderheads all around us. We read poems and delivered eulogies. The kids sprinkled flower petals over Simon's thawing form, which mercifully remained wrapped in a blanket. I added a tug toy for the afterlife.

What followed was the best version of a Southern-style celebration I could pull off in Vermont. We fortified ourselves with cold gin and told stories–like the time Simon ate an entire buffet-style taco dinner I'd set out for company. I returned from the bathroom to find the whole meal wiped out, right down to the bowls of shredded lettuce and jalapeños.

For the funeral lunch, my husband's aunt set out polished silver and her best china. A mourner, who happened to be a chef, provided edible succor.

The cake I wept into was moist.

Licked to Death by a Pit Bull

BY BRONWEN DICKEY

In the beginning, there was Angel.

I met her in the mountains of upstate South Carolina back in the winter of 2008—she belonged to some friends of mine—and the minute she trotted out to greet me, I felt certain that things would not go well. Her head looked like an anvil, for starters; it was framed by a wide jaw and lupine, almond-shaped eyes. Her silky black fur stretched over at least fifty pounds of muscle—she had the kind of physique you'd expect to see on a panther, not a pet. All the better to chase me down and devour me with, of course, because Angel was some kind of demon dog. You could tell just by looking at her. Angel was pure pit bull.

As it turned out, Angel wasn't much of a fighter; she was more of a leaner. Astonishingly obedient. A bit on the needy side, if you want to know the truth. When the time came for her to chase the horses back into their corral, she did her job like an old pro, with precision and care, but most of the time she seemed more interested in soaking up human affection, however she could get it.

So when the time came last year for my husband, Sean, and me to give our imperious, grumpy little pug some company, I started doing research on pit bulls, a breed I had always been taught to fear and revile. (Was I insane? Aren't they bred for blood? Don't they turn on their owners, and maul children without the slightest provocation? Don't they have locking jaws?!)

Well, no. And no, and no, and no, and no (no dog has locking jaws, by the way, and a pit bull's bite is weaker than, say, a German shepherd's). There is no real DNA profile for the "pit-bull-type dog"; it's at best a catchall term for what is pretty much a mutt all around, but I was shocked to learn that the American bulldog-terrier mix was actually once cherished as a national icon, the canine embodiment of loyalty and courage and rock-solid temperament. The kind of dog you could always count on, and the kind you could trust with any job, from cutting cattle to search and rescue to, yes, hanging out with the kids. Petey, the Little Rascals' sidekick from *Our Gang*? He was a pit bull. The RCA Victrola dog? A pit bull. The Buster Brown mascot? Pit bull. Sergeant Stubby, the most highly decorated dog in World War I? Pit bull. Portraits of pits draped in the American flag graced some of the most famous wartime recruitment posters. Even Theodore Roosevelt and Helen Keller adored pit bull dogs.

When I told friends what I had learned, they hemmed and hawed as though I were considering the acquisition of a Bengal tiger. One politely told me that she "assumed certain things about people who owned pit bulls." My mother, well versed in the child-mauling-locking-jaw spiel, claimed I had a death wish. But the scales had fallen from my eyes. If a pit bull had been good enough for Helen Keller, then–what the hell?–I figured one was damn well good enough for us. So we decided to take our chances with the most notorious dog in America. And we had no trouble at all finding one, because all our shelters in North Carolina seemed to be overflowing with them.

"Pits have a hard time here," one of the shelter volunteers told us, "because people are so scared of them. They're surrendered all the time in the worst possible shape–sick, starved, beaten, tortured, you name it. And we have to put a lot of them down, which is such a shame, because they make excellent family pets."

We selected a young tan-and-white female with a red nose and honey-colored eyes who bounded over to us like a gazelle the first time we met her. Sean and I had recently returned from New Orleans and the Saints had just won the Super Bowl; our new addition, an underdog if there ever was one, looked elegant yet tough, refined yet scrappy. What could we do but name her Nola?

In the first few months after bringing Nola home, she consistently surprised us in every way. Our "junkyard dog," she of dubious lineage and dangerous reputation, was more elaborate with her affection than any canine either of us had ever owned–more than all those retrievers, spaniels, hounds, terriers, and shepherds put together. If we were in any danger at all, it was the danger of having our faces licked off, the danger of drowning in slobber.

Wherever one of us went, Nola trundled alongside, and wherever we reclined together, Nola wedged between us like a balloon at a seventh-grade dance, curling into a bizarre contortion that we now call the "pit ball." She dutifully checked the perimeter of whatever room we happened to be in. She groomed us and nuzzled us and rolled onto her side to spoon when we watched movies. Since we could never seem to peel her off of us, I joked that we might as well put a bonnet on her and start pushing her around in a stroller. (When I was at home alone at night, however, I didn't exactly mind having a pit bull at my side. Potential intruders didn't need to know that she was a love sponge.)

Every time I looked at Nola, she dished her ears forward, cocked her head appraisingly, and furrowed her brow in a way that let me know gears were turning back there, trying desperately to figure out what I wanted her to do. If I took her out for a hilly three-mile trail run, she pushed herself to the limit, racing ahead like some kind of spotted bullet. If I felt under the weather, she was content resting her head in the crook of my arm while I read a book. She picked up new commands and solved puzzle toys in minutes (thanks for nothing, Kong!), so our main challenge, if you can call it that, was keeping her from being bored. To paraphrase the late animal behaviorist and pit bull advocate Vicki Hearne, it was as though we weren't so much training Nola as we were reminding her of something.

But when it came time to take her out in public, people reliably cringed and scooted away from Nola. I tried to offer up to wary strangers all the counterintuitive factoids I had come across from veterinarians and behaviorists–like the fact that pits are some of the most social dogs around, that they rank right up there with Labs and golden retrievers in terms of how much they seek out human attention. Or that the American Temperament Test Society, which has tested nearly nine hundred pit bulls, gives them a passing score of 87 percent, higher than that of beagles and border collies.

Even armed with the data, we quickly realized that Nola's affectionate nature was no match for decades of media hype. That didn't make me sad for her (she didn't know the difference) as much as it saddened me for the thousands of stable, adoptable pit-bull-type dogs in shelters across America that are euthanized every year because of this hysteria (we'll never know quite how many, because the phrase *pit bull* means different things to different people), and for the folks I met who were missing out on the companionship of such capable, versatile pets.

We have all read those headlines, hundreds of them, about horrifying, often fatal, pit bull attacks, and after Michael Vick's famous arrest, we are all more familiar than we probably want to be with the evils of the dogfighting industry. Fear sells much better than reason, but fear also can't bloom without ignorance.

Chain up any kind of dog, subject it to the jeers and taunts of passing strangers, and deny it food, shelter, and meaningful human company, and you may very well end up with a dangerous, unstable animal. With pit bulls, the media-stoked firestorm about their "viciousness" has created a tragic feedback loop: They have a terrible reputation, so reckless owners sometimes treat them miserably, then a subset of these dogs ends up reinforcing the stereotype–though many defy it regardless of their upbringing. It's important to remember that behind every broken dog is a severely broken person. You can't have one without the other.

In a hundred years, the pit bull has gone from national hero to unpredictable monster, and the dogs are still the same. We're the ones who have changed. Despite the variances in their size and shape and traditional uses, all breeds of the domesticated dog are part of the same subspecies: *Canis lupus familiaris*. The strongest element in their DNA is that they want to be with us, that they want to do what we ask of them. That is both the blessing and the burden of their loyalty.

As I write this, my arm is buckling under the significant weight of a big, blocky head. A pink nose is twitching near my keyboard, and every so often, a heaving sigh escapes it. I am being stared at with an intensity that tells me to please hurry up, it's way past dinnertime, waiting has now become unacceptable. So I will end with this: I now make certain assumptions about people who own pit bulls, too. I assume they are independent thinkers, they have transcended a long-standing prejudice, and, more important, they know a damn good dog when they see one.

Fetch Daddy a Drink

BY P. J. O'ROURKE

I have three badly behaved children and a damn good bird dog. My Brittany spaniel, Millie (age seven), is far more biddable and obedient than my daughters, Muffin (eleven) and Poppet (nine), and has a better nose than my son, Buster (five). Buster does smell, but in his case it's an intransitive verb.

My dog is perdition to the woodcock and ruffled grouse we hunt hereabouts and death itself to the pen-raised Huns, chukars, and quail she encounters at the local shooting club. Millie hunts close, quarters well, points beautifully, is staunch to wing and shot, and retrieves with verve. My children . . . are doing okay in school, I guess. They look very sweet—when they're asleep.

As my family was growing, I got a lot of excellent advice about discipline, responsibility, respect, affection, and cultivation of the work ethic. Unfortunately this advice was from dog trainers and was directed to my dog. In the matter of child rearing there was also plenty of advice, all of it contradictory–from family and family-in-law, wife, wife's girlfriends, pediatricians, nursery school teachers, babysitters, neighbors, and random old ladies on the street, plus Dr. Spock, Dr. Phil, and, for all I know, Dr Pepper: Spank them/Don't spank them. Make them clean their plates/ Keep them from overeating. Potty train them at one/Send them to Potty Training Camp at fourteen. Hover over their every activity/ Get out of their faces. Don't drink or smoke during pregnancy/ Junior colleges need students too. And none of this advice works when you're trying to get the kids to quit playing video games and go to bed.

It took me years to realize that I should stop asking myself what I'm doing wrong as a parent and start asking myself what I'm doing right as a dog handler.

The first right thing I do is read and reread *Gun Dog* by the late Richard A. Wolters. This is the book that revolutionized dog training in 1961. (Of course, the dogs are now fifty-three years old and not much use, but the book is still great.)

"Start 'em young" is the message from Wolters. And that's why, if we have another child, he's going to learn to walk pushing on the handle of a Toro in the yard instead of teetering along the edge of the sofa cushions in the living room. Wolters, along with a number of other bird shooters, had realized that waiting until the traditional one-year mark before teaching a puppy to hunt was like carrying your kid in a Snugli until he was seven. Wolters was sure he was right about this, but he wasn't sure why. Then he came across the work of Dr. John Paul Scott, a founder of the Animal Behavior Society. Dr. Scott was involved in a project to help Guide Dogs for the Blind. Seeing Eye dog training was considered almost too difficult to be worthwhile. Using litters from even the best bloodlines, the success rate for guide-dog training was only 20 percent. Dr. Scott discovered that if training began at five weeks instead of a year, and continued uninterrupted, the success rate rose to 90 percent.

It goes without saying that the idea of Seeing Eye kids is wrong–probably against child labor laws and an awful thing to do to blind people. But I take Dr. Scott's point. And so did Richard Wolters, who devised a gundog training regime that had dogs field-ready at as early as six months. That's three and a half in kid years. My kids weren't doing anything at three and a half, other than at night in their Pull-Ups.

The Start-'Em-Young program turns out to be a surprise blessing for dads. Wolters writes in *Gun Dog* of a puppy's first twenty-eight days (equal to about six months for a kid), "Removal from Mother at this time is drastic." That's just what I told my wife about the care and feeding of our infants–*drastic* is the word for leaving it to me. According to Wolters, I'm really not supposed to get involved until the kid is one (equivalent to a fifty-six-day-old pup). Then I can commence the nurturing (Happy Meals) and the "establishing rapport" (sitting with me on the couch watching football).

Next the training proper begins. "Repetition, more repetition, and still more repetition," enjoins Wolters. I've reached the age where I'm repeating myself all the time, so this is easy. "Commands should be short, brisk, single words: SIT, FETCH, WHOA, COME, NO, etc." In the case of my kids the "etc." will be GETAJOB or at least MARRYMONEY.

"Keep lessons short," writes Wolters. And that must be good

advice because notice how all the fancy private schools start later, end earlier, and get much more time off at Christmas and Easter than P.S. 1248. Wolters also points out that body language is important to the training process. "Your movements should be slow and deliberate, never quick and jerky." Martinis work for me.

"Don't clutter up his brain with useless nonsense," warns Wolters, who is opposed to tricks such as "roll over" or "play Dick Cheney's lawyer" for dogs that have a serious purpose in life. Therefore, no, Muffin, Poppet, and Buster, I am not paying your college tuition so you can take a course called "Post-Marxist Structuralism in Fantasy/Sci-Fi Film." And, meanwhile, no, you can't have a Wii either.

Wolters favors corporal punishment for deliberate disobedience. "Failure to discipline is crueler," he claims. I do not recall my own dad's failure to discipline as being crueler than his pants-seat handiwork, but that may be my failing memory. In any case, a whack on the hindquarters is a last resort. Wolters prefers to use psychology: "You can hurt a dog just as much by ignoring him. For example, if you're trying to teach him SIT and STAY, but he gets up and comes to you, *ignore him.*" When I was a kid, we called this Dad working late every day of the week and playing golf all Sunday.

According to Wolters, the basic commands for a gundog are SIT, STAY, COME, and WHOA. With no double entendre intended concerning the GIT OVER HERE directive, those are exactly the four things my boy Buster will have to learn if he wants a happy marriage. My girls Muffin and Poppet, on the other hand, seem to have arrived from the womb with a full understanding of these actions–and how to order everyone to do them.

"The last two, COME and WHOA," writes Wolters, "are so important that if a dog had good hunting instincts and knew only these two commands he would make a gundog." It's the same for accomplishment in every other field, among people and pooches alike. If you had to give just two rules for success in business, politics, family, friendship, or even church, you could do a lot worse than SHOW UP and SHUT UP.

Wolters begins, however, with SIT and STAY. And these are important too. Kids today are given frequent encouragement to STAND UP FOR THIS AND THAT. But SIT TIGHT 'TIL IT BLOWS OVER is wiser counsel. Wolters employs a leash to pull the head up as he pushes the rump down. I've found that the collar of a T-shirt works just as well. Wolters uses praise in the place of dog biscuits; he writes, "I do not believe in paying off a dog by shoving food into his mouth." I, on the other hand, try to make sure the kids eat their green leafy vegetables once I've got them seated.

Wolters teaches STAY by slowly moving away from the dog while repeating the command and making a hand signal with an upright palm. But I've found that if your kids get Nickelodeon on cable TV, you don't have to say or do anything. They'll stay right there in front of it for hours.

Once SIT and STAY have been mastered, you can go on to

COME. Wolters lowers his palm as a signal to go with the command, but a cell phone signal will also work if your kids are properly trained. Mine aren't. Getting a kid to come when he's called is a lot harder than getting a dog to, probably because the dog is almost certain that you don't have green leafy vegetables in the pocket of your shooting jacket. Wolters suggests that if you're having trouble teaching COME, you should run away, thereby enticing the dog to run after you. This has been tried with kids in divorce after divorce all across America, with mixed results.

The command that's the most fun to teach using Wolters's method is WHOA:

"The dog," writes Wolters, "is ready to learn WHOA as soon as he will STAY on hand signal alone and COME on command. When he has this down pat, my system is–*scare the hell out of the dog.* Put the pup in the SIT STAY position. Walk a good distance away from him. Command COME. Run like hell away from him. Make him get up steam. Then reverse your field. Turn, run at the dog. Shout WHOA. Thrust the hand up in the STAY hand signal like a traffic cop. Jump in the air at him. *Do it with gusto.* You'll look so foolish doing it that he'll stop."

Personally, I don't have to go to this much trouble. Just my morning appearance–hungover, unshaven, wearing my ratty bathrobe and slippers Millie chewed–is enough to stop my children cold. I reserve the antics that Wolters describes for commands to this idiot computer I'm writing on. *Gun Dog* was authored in the days of the simple, reliable Royal Portable. Thus Wolters has nothing to say about computers. Besides, dogs don't use computers. (Although, on my Visa bills, I've noticed some charges to rottenmeat.com.)

Children don't need computer training, either. Muffin, Poppet, and Buster–who can't even read–have "good computing instincts." When the Internet says COME, they come. Mom and Dad try WHOA on certain websites, but whether that works we can't tell. I'm the one who should be taught some basic commands, to make this darned PC . . .

"What's the matter, Daddy?" Muffin asks. With one deft flick of the mouse thing, she persuades the balky printer to disgorge all that I have composed. I see her frown. "Daddy, Millie chews everybody's shoes. She bit the teenager that mows the lawn. She killed Mom's chickens. And every time you come home from hunting, you're all red in the face and yelling that you're going to sell her to a Korean restaurant. And . . ."

And here is where my Richard A. Wolters theory of parenting goes to pieces. There is one crucial difference between children and dogs. You can teach a dog to lie. DOWN.

The World's Greatest Dog

BY ROBERT HICKS

Much of my career has been spent writing and saying things that fly in the face of conventional thinking. Whether explaining why I believe that fiction has a critical role in preserving history or even why it is progress to tear down a pizza shop to restore a long-lost battlefield—let's be honest, it's all been an uphill battle. But there are probably no words I have ever spoken that seem to draw a reaction, if not irritate folks, more than his name.

You see, my dog is named Jake, the World's Greatest Dog. That's his full name. Oh, he most often lets folks slip into the informal and simply call him Jake, but like the *tu* in French, you don't get there until you know him.

Fellow dog owners especially seem to chafe at this as if I had purposely set out to rouse their ire with his name. That is, until they meet him.

I usually have to wade through their counterclaims about Fifi or

Roger or Max. It's always pretty much the same. They, their family, have always believed that they have the World's Greatest Dog and on and on and on. Really, I sometimes wish that Jake wasn't the World's Greatest Dog just so I wouldn't have to hear about why Roger is. And believe me, before they finish I know all about Roger's pedigree, all his tricks, how he saved that child from the abyss and all the rest.

Now, don't get me wrong. There are a heck of a lot of really great dogs out there. I've met a whole slew of them over the years and even owned Banjo, Cindy, and Salem. Truth is, I'm of the school that a bad dog is pretty much nothing more than the victim of a bad owner.

So if you do own a really great dog, don't feel the need to stop me on the street or write me a letter telling me that I need to meet your Roger before I make claims I can't back up.

And while we're at it, let me clarify that Jake, the World's Greatest Dog doesn't come that way by some lofty German short-haired pointer pedigree dating back to the 1730s. No siree! As best DNA can explain, Jake is half Rhodesian ridgeback, one-eighth pit bull, one-eighth chow, one-eighth Lab, and one-eighth golden retriever. Or, as Jake and I would like to think, a perfect mix.

He looks most like his Lab ancestor, though I will admit he does have a bit of a ridge on his back and is chow in the ears. But his fur is pure Lab, which is a pity, as he really is all Rhodesian ridgeback in his heart and likes to envision himself on the Serengeti Plain, stalking the scent of lions ahead. In fact, while he never complains when you bathe him, his lack of protest is more because he's not a complainer than because he likes water. He simply stands there and takes it like the man he is.

Nor is he a retriever . . . of anything . . . well, other than the pelvis of a dead buck he found in the woods around my cabin near Bingham, Tennessee. But, even then, he wasn't "retrieving" as much as fulfilling his role as a lion hunter. Truth is, other than the simplest of commands to come, sit, or go to bed, Jake doesn't have any tricks up his paw.

So now that I have ruled out pedigree, tricks, and acts of heroism, you might be asking why the heck is Jake the World's Greatest Dog? But I know why you're asking that. It's because you have never met him.

You see, Jake has one overriding gift and that is the gift to love. Oh, I know your dog loves you back, but with Jake it seems different. All those folks who were telling me about why their dog could give Jake a run for the money start getting really quiet in his presence. No sooner is he there in your company than he turns it on. The love, that is. Whatever it takes. Usually his stare is enough to melt your heart. Lord knows, no one has ever accused Jake of not giving good eye contact. But, if that doesn't work, he will begin to lean into you as if to say I will not be ignored, you *will* love me.

He has yet to meet anyone whom he doesn't immediately take to. Even folks who don't like dogs seem to warm up in his company. Of course, all this makes for a lousy watchdog, but then again, I warned you, no tricks. In fact, the only time he ever barked, he scared himself as if he believed he was above barking and really didn't know he had it in him.

Jake was found on death row in a shelter in Columbia, Tennessee, by Michele Preston, a vet and the best friend a dog has ever had. He appeared to be about four years old. What she saw that day was an extraordinary dog–intelligent, kind, loyal. She gave him to Justin Stelter, a friend of mine, who has a real heart for dogs. They were together for several years until Justin married and moved to a place where he couldn't take Jake.

Justin came to me and made an offer I couldn't accept. He wanted me to take Jake for a year or so until they could find a better place to live and then give him back. I couldn't do it. Like anyone who has ever met Jake, I was already attached. The best I could do was to take Jake and promise to love him back from here on. Justin eventually said yes to my counteroffer, and Jake came home to my cabin.

In the years since, Jake has made quite a name for himself. He has appeared in a host of magazines and newspapers and even a few books. He was grand marshal of both the Franklin and Leiper's Fork Christmas Parades, and has pleased the hearts of a lot of folks, young and old alike.

If there is a downside to Jake, and I am hard-pressed to find one, it may be his passion for women's crotches. But even then how can I call him out on this? As an ancient grande dame around here said when I turned and found his nose buried deep in her crotch, and pulled on his leash, "At my age, I'll take what I can get."

But even after all these words, you still may not get the power that Jake has over the world. I could recount the time I was in New Orleans and Jake decided he was lonely and made his way back to my cabin from my friends who were keeping him, some seven miles away, in less than two hours, crossing highways and fording a small river, arriving torn up from a run-in with barbed wire. But, even then, lots of dogs have done far greater feats.

No, Jake remains the World's Greatest Dog simply because he loves. It's not enough that he has never met a man or woman, a fellow dog, a cat, a squirrel, a skunk, or a wild turkey he didn't love, it's the quality of his love for every living creature. And those who encounter him pick up on it quickly. As I said, it silences both the boasting of my fellow dog owners and those who find dogs annoying.

He wrinkles his forehead and begins to smile at them as he lets them know that no matter what they say or do, he is going to love them. My friend Marianne Schroar likes to tell me that Jake loves her more than he loves me. I don't argue with her, as I know that it's hard to be in his company without thinking that, somehow, surely, he must love me more than anyone else on earth.

So state your case as to why we need another pizza shop on hallowed ground or even why you find historic fiction confusing (Shakespeare will be joining me on that one), but be prepared to lose your argument about the World's Greatest Dog. For right there, next to me, will be Jake. He's been looking forward to meeting you.

Guide Dogs

BY RICK BASS

Since moving to Montana in 1987, I have owned only bird dogs. There are a lot of wild game birds in Montana, and I like to hunt and eat them. But the first dogs I had as an adult were not hunters, or rather, I did not use them as hunters, but as pets, and—like the opposite of Cerberus—as guides into a life of art, and a life of the mind.

I was working too much. I was putting in eighty-hour weeks—forty at the office and forty on the road, back and forth to the shallow oil and gas fields of North Mississippi and North Alabama. I spent my life, back when I was an oil geologist, dreaming of buried treasure, constructing intricate maps of worlds that lay just below us all, unseen and unknown, but deeply desired. I was just starting to hang out at the great bookstore in Jackson, too—Lemuria—where I was being hand-sold a couple or three books at a time, the staples of Southern literature, new and old. Eudora Welty, Barry

Hannah, William Faulkner. Willie Morris, Larry Brown, Reynolds Price, Robert Penn Warren, Flannery O'Connor.

The short-story renaissance–at least the second one–was going on then, too; great new stories were appearing every week, not just in the major magazines but in the explosion of independent literary journals. And there was something particularly wakeful in me, particularly mindful, to realize that these were not stories from the long ago, but instead those for which the ink was barely dry. Stories that were intensely of our time, so much so that there was very much the feeling that the next week, next month, next paragraph would bring more insight and even instruction on how to navigate the slumber of those strange times, the eighties.

Back then, in Mississippi, one day at dusk I was driving out into the country to catch up with my girlfriend. She lived along the bayou, in a farmhouse at the edge of a vast field of soybeans and cotton. It was like driving back into another century, and another country, to get there–up and down the ridges and ravines, driving with the windows down, through cold pockets and lenses of air as warm as breath. I drove past two scrawny little black-and-tan pups roadside, probably dumped there. They sat piteously next to a third pup, which was dead. I drove on past, wincing. I went another mile down the road before turning around. I went back and picked them up. I didn't mean to keep them–I was going to turn them in to the animal shelter the next day–but it just didn't work out that way. I kept them in a shoe box that first night–how could they ever have been so small?–and gave them some milk.

As with most siblings, the two females occupied different territories, sometimes even different universes. Ann was a little bull, a ruffian, never saw a mud puddle she wouldn't walk through, while

Homer, though similarly marked, was slender, a wisp. Walking her on a leash was like having a helium balloon at the end of the tether. They were the two most well-behaved dogs, never failed to heed a single command the first time. I like to think they were so grateful for having been rescued that each day—ten good years for Ann, and fourteen for Homer—they never forgot how things could have gone; that they were each delighted simply to be alive, and to be a dog. Homer never disobeyed a day in her life. I wonder now if their distinct personalities didn't somehow allow me to better develop as a writer, or at a pace better suited for me—slowly, organically—by enabling me to inhabit, particularly across those early years, a household suddenly filled with two such separate types: the pensive, delicate, loyal Homer, and the gluttonous, devil-may-care Ann. As if in caring for them, I had to use, and grow, both halves of my brain, simultaneously and equally. Who knows? The facts are indisputable. I picked them up, and good things began happening to me.

Within a few short months I had quit my day job, become a writer, won some awards, moved west, met an editor who loved those dogs and agreed to read my stories and became my publisher. My girlfriend had made the move with me; we later married and had children, went for long hikes in the mountains with those old black-and-tan hounds, had many excellent adventures, lived happily ever after. The hounds accommodated every change that came to them, became proud guardians of our first daughter and then our second. On our camping trips, the inside of our tent was a nest of sleeping bags, the two dogs and the two girls all about the same size, and all snarled up in a pack, sleeping soundly.

It would have been easier to keep on driving, and let someone else do the good work of stopping to pick the pups up. And maybe all that good luck would have gone somewhere else, or not have happened at all. Everything happened, it seemed, from that small kindness. My life cracked open. It had not been a bad life–had in fact been pretty good–but in some ways it was as if I had been sleeping. A lot of young people are prone sometimes to think perhaps a little overmuch about themselves.

I think what they opened my eyes to were kindness and vulnerability. To hold each pup in one hand–it sounds corny, but again, I remember it as if it were yesterday–something opened, something awakened. It was a small thing, but that was long ago and it has not yet closed, though those old hounds are deep beneath our feet now, buried, like so many other treasures.

Sit, Stay, Heal

BY JACK HITT

What kind of day was it? The morning began with an oncologist. She laid out our next year—my wife's cancer would require radiation, then chemo, then an operation, then more chemo. Meanwhile, my dog, Jesse, was run over by a car. It was that kind of day.

I was sitting in an uncomfortable chair in a bland office, struggling to comprehend large, ugly words, when my phone hummed in my pocket. I slyly checked and saw that it was my friend Tom, who was painting my house. I clicked ignore and tuned back in.

Right, my wife had cancer. And this was the talk—the big talk—where the doctor tells you how it's going to be. So many words—*adenocarcinoma*, *endoscopic*, *lymphatic*—words that held little meaning for me, just tiny sonic pricks of dread.

The phone hummed once more. Again, it was Tom—click. *Radiation scarring*, *metastasize*—they floated in the air like jazzy verbal riffs, Charlie Brown's teacher whose voice is a muffled trombone. Again, the phone: Mmmm. Mmmm. Again, click. *Leucovorin, adjuvant, epithelial.* The doctor's voice resurfaced into standard English long enough for me to hear "a winter of nausea." Mmmm. Mmmm.

"You know, this is crazy, but Tom has called me four times," I said. "Maybe my house is on fire? Excuse me a minute." I stepped outside of the office to hear Tom, very upset, explain that Jesse had bolted out the front door and had been run over by a car. His pelvis was crushed and he lay in the street howling. The driver freaked out and, howling also, sped off. Tom slid Jesse onto a piece of plywood, loaded him into the pickup, and drove like a bat out of hell to the animal hospital.

"Mr. Hitt, your dog is severely injured," the vet said, taking the phone from Tom.

"I see. Look, I am in an oncologist's office right now. Here's my question. Will my dog be alive in fifteen minutes?"

"Mr. Hitt, this is very serious," she said. "You really need to get here right now!" We had this exchange two more times with the vet getting increasingly agitated with me.

"You need to deal with this right now! Mr. Hitt, your dog has suffered a broken pelvis, contusions on the–"

"Let me try this one more time, ma'am. I am in an *oncologist's* office. Do you know what that word means?" A gut-wrenching silence seized the phone. For the first time, I felt the awesome power of illness, my first time at playing the C card. It was, paradoxically, invigorating.

"Fifteen minutes," she said. "Yes, we'll be here."

We had gotten Jesse some six years before when my two daughters, in their late single digits, promised they would feed and walk him every day. (I was so young and naive back then.) After giving a stern daddy lecture about rational choice and how we were going to visit lots of kennels and shelters, we never got past the first puppy at the first stop. He was a ball of ebony cashmere who could curl into a perfect sphere of drowsy euphoria. The kids immediately pronounced his name, and we brought him home to an orgy of affection. Jesse seemed, at times, a canine genius, and at others, hilariously challenged. In a few days, he was perfectly housebroken, but he struggled with the basic concepts of "stay" and "come" and "heel."

We even took him to obedience school. It was in a warehouse and all the owners would stand in a formation with their dogs panting attentively at their sides. Jesse wandered about, as if in a verdant pasture, deaf to all instructions issued in every possible tone of voice—ranging from pleading falsetto to enraged Parris Island drill instructor. Finally, he'd locate some distant scaffolding-filled corner and just plop down like Ferdinand the bull.

When he was two years old, though, he somehow awoke to the pleasures of cooperation. He began to respond to commands and actually seemed to take great pleasure in being your best friend.

And he grew into a big beautiful dog. He developed some brown on his paws and underbelly and some black on his tongue. Our little mutt revealed himself to be part Rottweiler and, probably, chow. He's muscular, with a low center of gravity and, if you don't know him, a menacing look. He has an introductory bark that can rattle china off a shelf. But when he cocks his head with half-bent ears as if staring into the abyss of an RCA Victrola, or when he gallops at you like a startled camel in a halo of slobber, it's impossible not to love the guy. He's a cute love bunny who just happens to look like the evil spawn of Cerberus.

At dog parks or in chance encounters, Jesse's a cheap hump and totally GGG, thrilled for a ride and happy to assume the position. He's insanely inquisitive, and takes it as his life mission to chase down any nearby rodent and bark at it before looking back at me, as if to say, "I got your back, boss." Being run over by a car was a kind of blessing. Lisa and I love this dog, and focusing on his life-or-death crisis made our affairs seem languidly manageable (which ultimately they were). On the way over to the animal hospital, I called my brother and sister, both crazy dog people, and sought their advice. Big dogs don't do well with broken bones, I was told, and give him a day or so and then decide if you need to put him down. A hard decision, they told me, but, sometimes, the only humane choice.

The vet, a sober brunette in her thirties, ushered us into the intensive care unit, where Jesse lay knocked out on a gurney, his paw shaved to the skin, an IV snaking out of his shank and into a bottle on a hook. He looked dead, a preparation for the decision we might have to make.

The vet asked us to step into another room to see the X-ray she had taken. This was the talk–the big talk–where the doctor tells you how it's going to be: that we needed to keep Jesse in the unit overnight. He might well die. She showed us the very image of his crushed pelvis. She explained how very soon, within a few days, we'd need to schedule major surgery. She said, "There is no other option."

People of Scottish descent, according to the ethnic slur, are notoriously miserly, and my mom's family is nothing but Highlanders all the way back. So, in the midst of all this breathless talk, I asked the rude question. How much am I in for so far? She said, "Seven hundred dollars." And for the overnight stay in the ICU? "Fourteen hundred dollars more"–although it was clear from her tone that she considered my questions vulgar. And for the surgery? "Twelve thousand dollars."

The whole room seemed to stand still. Most of us don't carry animal health-care insurance, so veterinarianism is a fee-for-service business.

Again she said, "There is no other option."

I now bore down into her eyes. "Here's the bad news," I said. "Where I grew up, there is another option." I could see she understood what I was suggesting, but she cocked her head, as if she were seeing the real me for the first time–a Victrolan vortex of evil. "I can put Jesse down."

She could not look at me, as I had now morphed into Heinric Himmler. That transformation was cemented the next minu when I coldly insisted that I wanted my dog. Now.

I paid the bill, situated Jesse into the sling of a blanket, placed him in the back of the car. At home, he lay half curl day on the kitchen floor, motionless, a black crescent of susp foreboding. By sundown, Jesse was awake but unable to m contact. At bedtime, I headed upstairs and Lisa said she' When I came down the next morning for coffee, I found sleeping on the linoleum beside our pup marooned o blankets.

He peed red that morning and I sped off in the car doggy downers. We narcotized Jesse silly those next then carried him and his blankets out to the backy temperate days and nights made it easier for him. One morning he sipped some water, and he looked me in the eye. I choked up. Even still, if you petted him at all, he'd grip your hand with his teeth. Jesse was never a biter and even in agony, he still wasn't. He'd just put his teeth on your hand to let you know: Don't even think about touching my hip.

At times he'd drag himself off the blanket and make a bit of a pee. I'd gently pick him up—his teeth on my hand the whole time—d put him back on the blanket.

ne afternoon, about ten days after the accident, I was stand-
the porch talking on my flip phone when a squirrel bolted
he yard. Jesse scrambled up on three trembling legs and
of his legendary barks. It was the first time I'd heard it
o weeks. The squirrel was gone as Jesse looked back
der, to let me know I was safe. I snapped the phone
nto tears.

put down animal hospitals, but I suspect he's
is current pelvis than any trussed with steel
He's back to his old self. He sprints like a

galumphs, and by streaking, I mean
I mean warthog. Jesse totally recov-
igh-minded of me not to mention
d dollars richer. Good dog.

Who's the Boss?

BY CHRIS HASTINGS

When it came to selecting my first gundog, I had one rule—and it came from my wife, Idie: "Bring home a female. There's too much testosterone in this house as is."

Truth be told, I needed the advice. I am the son of an architect and an artist, so hunting and guns were not a regular part of my Southern upbringing. When I arrived in Birmingham, Alabama, in 1986 to work with Frank Stitt at Highlands Bar and Grill, I had yet to be exposed to quail hunting and the traditions and passions it inspires. And I certainly had never considered what it meant to own a bird dog.

After a few trips in the fields, I quickly decided that those who owned and trained their own gundogs were enjoying the apex of the wing-shooting lifestyle. When you see a man and dog in action it's damn near impossible not to want to experience it for yourself. So in 2001, after sixteen years of following others' dogs through the woods, I decided to join the gundog fraternity. Now I just needed to find the right dog.

Over the years I had the privilege of traveling to a number of great hunting camps—not because I am a particularly great shot or a brilliant conversationalist, but rather because I can cook a decent camp meal. During that time, I began to learn what it takes to find, train, work with, and hunt with a gundog. I was given names of kennels, countless dog-training books, and tons of advice. A whole world of information began to flow from people I had never met but who had heard I was in the market for a puppy, specifically a setter puppy.

But my biggest break came when I ran into Charlie Perry. Perry is an avid outdoorsman and dog owner but also, most important for me, a dog swapper—a specialized dog man who curries favor with the best trainers, breeders, and plantation owners in the country in order to have access to puppies. There are lots of these guys out there, some better than others. Perry makes an art of it. For our deal, Perry wanted the right to hunt with the dog when he desired and to have a pick of her first litter.

Perry told me that Mrs. John Harbert, owner of the famous Pinebloom Plantation near Albany, Georgia, owed him a puppy, and asked if I would like to have that dog. Larry Moon had run Pinebloom's kennel for years and earned a reputation for producing some of the finest bird dogs in the country, many of them regularly winning national field trials. To say your dog comes from Pinebloom gives you instant credibility. It was my shot at the dog of a lifetime. So off I went to South Georgia with the aforementioned set of instructions from my wife—no males.

I met Moon at the kennel, and he took me to see the pups. They were sixteen weeks old and had already started getting exposed to the outside world. When it came to choosing my puppy, there wasn't really an aha moment. I simply asked about the two females and Moon pointed to the one with "the most hunt in her." That was all I needed. I named her Sadie, put her in the car, and headed back to Birmingham.

Having read a number of books on training and spoken with every person I knew and respected in the dog-training world, I set about the task of first bonding with and then attempting to train Sadie. The daylight hours before my shift in the kitchen were too hot for training sessions. So for the first few months we trained after work, between 11:00 p.m. and midnight, at a local golf course. There we'd attempt basic yard commands–how to heel, come, and respond to a whistle. Our first night out was pretty telling of our future together. Having gathered my training gear, I made the mistake of cracking the door before clipping the check cord to Sadie's collar. Away she went. It took me thirty minutes to find her.

At the time, I was forty years old. Having driven my wife, my children, my staff, and a number of my friends half crazy with my overbearing attitude and insistence on things being done a certain way, I was ready to settle into less of an alpha personality. But Sadie gladly assumed that role. The first thing you notice in an alpha dog is that it would rather strike off alone than be with you. And when you give a command, it may or may not pay attention, depending on its mood. In many ways, training an alpha dog is not unlike raising teenagers. In the field, an alpha bird dog will follow its nose into the next county, paying little heed to commands to stop. And as you may have guessed, it's tricky to shoot a quail when your dog goes on point a mile from where you're standing. Cullom Walker, a dear friend and father figure, once told me, "I don't hunt dogs, I hunt birds. If the dog doesn't want to be with us, fine. We have other dogs."

Eventually, I began to get advice from those who had traveled down the same road I was on. At first they treated me gently, speaking in the third person about someone they'd known who would always yell at his dog, or who would constantly whistle for the dog to come, or who was perhaps a little quick to reprimand her. Then they began to speak directly to me, *about* me. Yet again, my wife offered sage advice: "Don't make us duct tape your mouth. Leave the dog alone." In other words, I needed to shut up and back off. It seems Sadie wasn't the only one displaying a bit of alpha.

As the years passed, Sadie and I both began to mellow out. She would show signs of affection and would even stick with me on hunts for extended periods of time. After five years, we both realized we didn't have to fight each other. Sadie is now nine years old, and I am forty-nine. She and I hunt beautifully together. I guess it comes down to this: We understand each other. I afford her an occasional disappearing act, but I now know she will return, or I will find her beyond my range of sight but waiting patiently on point.

How to Name a Dog

BY DANIEL WALLACE

Mugsy was a boxer. Resonating within that sentence is the simplicity of the perfect name. The same words appear to mean two completely different things. They could mean *Mugsy was a breed of dog called a boxer.* But I could also be referring to a man named Mugsy whose profession was that of *being* a boxer, because Mugsy is the kind of name we would expect a boxer to have—Mugsy, Rocky, Jake. That sort of thing. The name Mugsy works because a boxer *looks* like a boxer, and in that sense it's easy to imagine what a dog like that might be named. One could even claim it's clichéd, but I think the only person who would claim that is the kind of person who would begin a sentence with the words *one could.*

I was in single digits when we had Mugsy. Mugsy chased cars with a joyful, indefatigable single-mindedness. Lots of dogs do, of course, but Mugsy actually *caught* them and clung to their still-turning tires with teeth that somehow never seemed the worse for wear. He wasn't hit so much as he was slammed into the road. Repeatedly. He'd wander back home all bloody smiles. I was heartbroken when I discovered he'd been sent away to live on an old lady's farm out in the country . . . where I believe he is to this day.

The first dog I named myself was Barney. Barney was a basset hound. Mugsy and Barney are names that operate within the same blatantly descriptive universe, I think. Barney, like most bassets, was a sad-looking dog, a dog with a worried expression, as if he were beset by constant troubles when the reality was that he was cared for and fed free of charge. He'd been fixed. He didn't have a worry in the world, but you wouldn't know that by looking at him. Barney is the name we ascribe to a sad man, the difference being that people aren't born sad, but if they're given this name out of the womb they will without a doubt *become* sad. The name dictates the sadness to follow. Dogs benefit from being dogs in that we have a good idea of what they'll look like and the general characteristics they possess before we give them their names. Naming dogs is a kind of blessing, an affirmation; naming people can be a curse.

One day my sister and I took Barney to a friend's house, and he wandered out into the woods and disappeared. How far and how fast could a basset hound go? Far enough and fast enough to disappear. We never saw him again.

Rudy may have been the most difficult dog of all to name because he didn't appear to be a dog. Unlike the first two dogs, Rudy was a mixed breed. He looked like he needed more time in the oven. He was that unfortunate combination of canine genetics that ends up–like kids mixing together fifty things they find in the refrigerator–not being much of anything at all.

His big red eyes were so needy, so pitiful, and when he looked at you, it was not love you saw but the last hopeless look of a man falling off a cliff. *Maybe you'll throw me a rope or something? Maybe? No? That's fine. I didn't expect you to.* He whimpered. He whined. He shivered for no good reason. Women seemed to like Rudy, but it was really just pity. My father hated him. Whenever the poor dog came into the room, he raised a magazine rolled in his hand like a club, threatening, for no reason I could fathom.

I nailed this name. Rudy worked for him. It was perfect.

Then one day he went to live on that same old lady's farm. That old lady really liked dogs.

When I was fifteen, I got my first real job, working at a veterinary hospital. The guy who owned the hospital bred black-and-tan coonhounds. The summer I worked there a litter was born, and I bought one of the puppies.

After the first three dogs and their generic names, I wanted something different. Something bright and original. The name I came up with was Colonel Mosby. John Singleton Mosby (1833–1916), also known as the Gray Ghost, was a Confederate Partisan Ranger, a guerrilla fighter. He was noted for his ability to go behind enemy lines, do something heroic, then elude his pursuers. This had nothing to do with the dog, of course, but I liked the sound of it: *Mosby*. It was distinguished, in an odd way, and that's what Mosby was. Distinguished and odd. Mugsy, Barney, and Rudy were really family dogs, but nobody else could claim Mosby, this elegant and goofy beautiful life. We were the picture of a boy and his dog.

He died, run over after slipping his chain while I was at school, the day before his graduation from obedience school. I found him in the tall grass on the side of the road, carried him home in my arms, and buried him in the backyard. The next day my mother wrote me a sweet card. All this time through all these dogs she'd been trying to protect me from the reality of death, and then I saw it, discovered it in the tall grass. There were five twenty-dollar bills in the card, as if in recompense. She said this was the worst thing that would ever happen to me.

It wasn't.

I was in college in Atlanta before I gave another dog a try. Orsin was an English bulldog, the runt of a litter of six, and claimed that name because of it. Little Orsin, we called him.

Orsin and I shared an apartment with friends who, though wonderful, had no air conditioning. It was a hot summer. Orsin spent much of his time sleeping in his water bowl. I'd wake up in the middle of the night and take a bath.

Through the next two years of temporary romances and passing friendships, Orsin was the only constant in my life. We slept together, ate together, were rarely apart. Then I got a job in Japan and had to leave Orsin with my sister. He died before I came back–an aneurysm, they told me. I thought that only happened to people. But after you have a few dogs, you learn: Everything that can happen to people can happen to dogs as well. And they do.

Laura and I were married on September 20, 2001. We'd been going out for a couple of years before that, and though we decided not to have kids of our own, getting a dog wasn't out of the question at all, so we went to the Animal Protection Society to look around for one. It was so sad, all those dogs staring at us as we walked by their sad cages, some of them barking, some growling, some retreating to a corner to whimper. These were dogs you knew no one would ever choose; they would never leave this place, and if any of them had names—and some of them did—they might never hear them again for the rest of their lives.

Polly was different. She was about twelve weeks old, a mix of two mystery breeds (though she may have had a little chow in her). She played it cool. She wasn't over-the-top eager to get out of there, but neither was she altogether shy. She waited until we stopped at her cage and then, somewhat demurely, walked to the wires to get a good look at us. We felt as if we were being studied, and chosen, as much as we were studying and choosing her.

Why Polly? Laura's sister's name is Molly, my sister's name is Holly, and people sometimes call Laura Lolly. So Polly was Polly because she was one of the girls. So sweet she never barked, and she liked to sleep as much as we did. Laura taught her to whisper, the only dog I've ever heard of who could do that.

She didn't live very long. It turned out she had arthritis, hip dysplasia, and an autoimmune disease. It's been four years and we're still finding her toys beneath living room chairs, and small pockets of her fine white fur nestled in a corner, like a bed for something very, very small.

Dogs have been hanging out with people for over ten thousand years. They are empty vessels we fill with a reflection of ourselves; or, alternatively, they come ready-made with their own strong personalities, which, insane as they sometimes are, we accept, because they accept ours. Having a dog is possessing a life, and dogs are in fact like children, but better, because they don't grow up to rob banks or hate you. They love you the same until they die.

For the dog we have now, a stuffed-animal-look-alike cockapoo, it took us a long time to come up with the right name. It was Hurley for a while, but that didn't work, and then it was Charlie. Still, we knew that wasn't his name. His name, it turned out, is Jasper. That's what we call him, and when we call him, he comes. Sometimes.

Acknowledgments

The readers of *Garden & Gun* deserve much credit for this book's existence. They've been asking for it since the magazine's first year. I hope it's all that they've wished for.

The editor of an anthology is only as good as the writers who contribute to the book. In my case, the writers of these pieces made my job a delight. I'm also fortunate to work with Donna Levine, copy chief of *Garden & Gun*. She embraced this book project with aplomb, tackling everything from commissioning to editing the essays. *G&G*'s deputy editor, David Mezz, also provided his usual wisdom and attention to detail. Art director Marshall McKinney secured the talents of Clint Hansen, whose smart and striking illustrations grace these pages.

The magazine's agent, Amy Hughes of Dunow, Carlson & Lerner Literary Agency, championed this idea from the start and found a great home for it with HarperWave. There Karen Rinaldi and Julie Will challenged us for the best, and made sure we delivered it.

G&G brand development director Jessica Hundhausen Der-rick and director of corporate communications Sterling Eason brought their talents to bear, making sure the book received its due attention around the South and beyond.

A big thank-you goes to the owners of *G&G*, Rebecca Wesson Darwin, Pierre Manigault, and Edward Bell III, for their unwavering dedication—and dog-friendly office policy.

Closer to home, well, literally, *at* home, my wife, Jenny, tirelessly offers support and counsel. She is my true north.

Finally, a long scratch behind the ears to the dogs, past and present—my own, Flap Jack, Salty, and Pritchard, and those everywhere. They give us everything they have and in doing so make us better.

About the Contributors

ACE ATKINS is the *New York Times* best-selling author of more than a dozen novels. A native of Alabama, Atkins once played SEC football and was nominated for a Pulitzer Prize on the newspaper-crime beat. He and his family and many animals live in Oxford, Mississippi.

RICK BASS is the author of more than thirty books of fiction and nonfiction, most recently the novel *All the Land to Hold Us*. He lives in northwest Montana, where he is a board member of the Yaak Valley Forest Council.

MARY LOU BENDRICK is the author of *Eat Where You Live: How to Find and Enjoy Fantastic Local and Sustainable Food No Matter Where You Live*. She has written for grist.org, *Orion*, the *Boston Globe*, *Whole Life Times*, and other publications. She currently works in voice-over and has a novel in the works.

ROY BLOUNT JR.'s twenty-three books include *Alphabetter Juice: Or, The Joy of Text* and *Long Time Leaving: Dispatches from Up South*. He is a *Garden & Gun* columnist, a panelist on *Wait*

Wait . . . Don't Tell Me!, and a member of the Fellowship of Southern Authors.

John Ed Bradley is the author of seven novels, including *Tupelo Nights* and *Call Me by My Name. Sports Illustrated* named his memoir, *It Never Rains in Tiger Stadium*, the Best Sports Book of the Year in 2007. Bradley is a former staff writer for the *Washington Post.* His books have been translated into seven languages, and his magazine stories have been widely anthologized, appearing in *The Best American Sports Writing, The Greatest Football Stories Ever Told*, and *Sports Illustrated: Fifty Years of Great Writing,* among other publications. Bradley grew up in Opelousas, Louisiana. He was an All-State football player in high school and a four-year letterman at LSU. In 1979, his senior year, he was a team captain and named All-SEC as well as Academic All-SEC. He lives today in Mandeville, Louisiana, with his wife, Kimberly, and daughter, Hannah.

Rick Bragg is the best-selling author of *All Over But the Shoutin'*, *Ava's Man, The Prince of Frogtown*, and other books, mostly on the working-class people of the American South. He is a winner of the Pulitzer Prize and many other writing awards, and is currently the Clarence Cason Professor of Writing at the University of Alabama. He is the worst fisherman on God's earth, and has cried over many dogs.

Nic Brown is the author of the story collection *Floodmarkers* and the novels *Doubles* and *Life Drawing.* His stories and essays have appeared in many publications, including the *New York Times*, the *Harvard Review*, and *Garden & Gun.*

DOMINIQUE BROWNING is the author of several books, most recently *Slow Love: How I Lost My Job, Put on My Pajamas, and Found Happiness*; she blogs at slowlovelife.com. She is the cofounder and senior director of Moms Clean Air Force, a special project of the Environmental Defense Fund. She was previously the editor in chief of Condé Nast's *House & Garden*; her magazine career includes positions at *Newsweek*, *Texas Monthly*, and *Esquire*.

C. J. CHIVERS is a senior writer for the *New York Times*, covering conflict, crime, the arms trade, and human rights. He is the author of *The Gun*, a history of automatic arms and their influence on human security and war. From 1988 until 1994, he was an infantry officer in the United States Marine Corps. He lives with his wife, Suzanne Keating, and their five children in New England, where they fish, read, and raise as much of their own food as they can.

KATIE CROUCH is the author of the novels *Girls in Trucks*, *Men and Dogs*, and *Abroad*. She lives in San Francisco with her family and Bernese mountain dog, Daisy.

JIM DEES is the host of the Oxford, Mississippi-based *Thacker Mountain Radio*, a music and literature program on Mississippi Public Broadcasting.

DAVID DIBENEDETTO is the editor in chief of *Garden & Gun*, where he oversees all media platforms. He's also the author of *On the Run: An Angler's Journey Down the Striper Coast*.

BRONWEN DICKEY is a contributing editor at *Oxford American*. She is currently working on her first book, a cultural history of pit bull dogs, which will be published by Alfred A. Knopf in 2015.

CLYDE EDGERTON is the author of ten novels, including *Walking Across Egypt* and *The Night Train*, and two nonfiction books. His most recent book is *Papadaddy's Book for New Fathers*. Edgerton is the Thomas S. Kenan III Distinguished Professor of Creative Writing at UNC-Wilmington and is a member of the Fellowship of Southern Writers. He and his family are proud to count among themselves a rescued dog, Addie—named in honor of Addie Bundren.

CHARLES GAINES is the author of *Stay Hungry* and *Pumping Iron*, both adapted into feature films. An avid outdoorsman, he also codeveloped the sport of paintball.

CHARLIE GEER is the author of the novel *Outbound: The Curious Secession of Latter-Day Charleston*, which won an Independent Publishers Award for regional fiction. Geer's work has appeared in the *Southern Review*, *Tin House*, the *Sun*, and other publications. He lives in southern Spain.

ALLISON GLOCK is an award-winning writer, a contributing editor for *Garden & Gun*, and a pit bull advocate. Her most recent book is *Changers*, the first of a four-part young adult novel series aimed at cultivating empathy in teens.

SUSAN GREGG GILMORE is the author of three novels, including *The Funeral Dress* and *Looking for Salvation at the Dairy Queen*. She lives in Chattanooga with her husband and three dogs, Bubba, Genevieve, and Otis.

VANESSA GREGORY's work has appeared in *Garden & Gun*, *Harper's*, and the *New York Times*, among other publications. She also teaches journalism at the University of Mississippi.

CHRIS HASTINGS is the chef at Birmingham's Hot and Hot Fish Club, which he owns with his wife, Idie. His first cookbook, *The Hot and Hot Fish Club Cookbook: A Celebration of Food, Family, and Traditions*, was published in 2009. He was awarded Best Chef: South by the James Beard Foundation in May 2012. He lives in Birmingham with Idie and their two sons, Zeb and Vincent.

BETH HATCHER is a freelance writer living in central North Carolina. A native Tar Heel, she loves good North Carolina barbecue (eastern style, of course), random road trips, and writing about the complexities of the *new* New South. Her writing has appeared in newspapers and magazines throughout the region.

ROBERT HICKS is the author of the best-selling books *The Widow of the South* and *A Separate Country*. A collector and preservationist, he lives in Tennessee and is currently working on his third novel.

JACK HITT is a contributing writer to the *New York Times Magazine*, as well as a contributor to *Garden & Gun*. His book, *Off the Road: A Modern-Day Walk Down the Pilgrim's Route into Spain*, was made into a 2011 motion picture, *The Way*, directed by Emilio Estevez and starring Martin Sheen. He has won the Livingston and Pope Awards, and most recently, his *Harper's* report on American anthropology was selected for a collection of the best science writing of the past twenty-five years, *The Best of the Best of American Science Writing*. His work also appears in the *New Yorker*, *Rolling Stone*, and *Wired*.

BLAIR HOBBS is a lecturer at the University of Mississippi, where she teaches poetry workshops. She is also a multimedia artist. She lives in Oxford, Mississippi, with her husband, John T. Edge, and son, Jess.

LISA HOWORTH lives in Oxford, Mississippi. Puppy Sal and one of Sal's sons, the Quarter Pounder, appear in her novel, *Flying Shoes.*

JEFF HULL is a freelance writer living in the Ninemile Valley in Montana. He is the author of the novel *Pale Morning Done* and the essay collection *Streams of Consciousness.* He is currently, sadly, dogless.

PAT JORDAN has been a freelance writer since 1968. He currently lives in Abbeville, South Carolina, with his three Shiba Inus, one Retraniel (retriever-spaniel rescue dog), and a parrot named Florence, after his mother . . . oh, and a wife, too.

HUNTER KENNEDY was born and raised among the canines prowling Columbia, South Carolina. His writing has appeared in the *Minus Times, T Magazine,* and *Garden & Gun,* where he is a contributing editor.

DONNA LEVINE is the copy chief of *Garden & Gun.* She also writes poetry under the name Donna Levine Gershon. She lives in Oxford, Mississippi, with her husband, their two daughters, and their cockapoo.

BETH MACY is an award-winning journalist based in Roanoke, Virginia. A 2010 Nieman Fellow in journalism at Harvard, she is the author of *Factory Man: How One Furniture Maker Battled Offshoring, Stayed Local—and Helped Save an American Town.*

GUY MARTIN, an Alabama native, is a contributing editor for *Garden & Gun* and *Condé Nast Traveler.* He has written for many other publications as well, including *Wired* (UK), the (London) *Sunday Telegraph,* the *Observer,* and the *New Yorker.* He writes about the South and about Eastern Europe, where he is at work on a book about the Cold War in Berlin. He lives in New York City, Alabama, and Berlin, spending much of his time on the many different airplanes that fly between them.

J. M. MARTIN lives with his family in New York City, where he writes, helps support artist residencies at the MacDowell Colony, and pines for the South every living minute of every day.

JILL MCCORKLE'S novels and short stories have garnered many awards, including the John Dos Passos Prize for Excellence in Literature and the North Carolina Award for Literature. She teaches creative writing in the MFA Program at NC State University. She is also a faculty member of the Bennington College Writing Seminars and a frequent instructor in the Sewanee Summer Writers Program.

BOB MCDILL is a retired songwriter and music publisher. He has written thirty-one number-one songs as well as songs for movies and television. He was elected to the Nashville Songwriters Hall of Fame in 1985 and is the only Nashville songwriter ever to be voted Writer of the Year by both Broadcast Music Inc. and the American Society of Composers, Authors and Publishers. In 2012 ASCAP presented McDill with the Golden Note Award to commemorate his "extraordinary place in American popular music." He now works as a freelance magazine writer.

THOMAS MCINTYRE has authored thousands of magazine articles and television scripts, and is the author of such critically acclaimed books as *Days Afield*, *Dreaming the Lion*, and *Seasons & Days*. His novel, *The Snow Leopard's Tale*, is a story of the Tibetan Plateau. His wife, Elaine, and he are currently being held hostage in their own home by an English cocker spaniel pup.

Bucky McMahon is a painter, sculptor, and nationally known travel and adventure writer.

Jon Meacham has written books about Franklin Roosevelt, Winston Churchill, Andrew Jackson, and Thomas Jefferson. An executive editor at Random House, he also teaches at Vanderbilt and at Sewanee.

Jonathan Miles is the author of *Dear American Airlines*, which was named a *New York Times* Notable Book and a Best Book of 2008 by the *Wall Street Journal* and the *Los Angeles Times*. He is a former columnist for the *New York Times*, and his journalism, essays, and literary criticism have appeared in the *New York Times Book Review*, *GQ*, *Details*, *Men's Journal*, the *New York Observer*, *Field & Stream*, *Outside*, *Garden & Gun*, *Food & Wine*, and many other magazines. His work has been included numerous times in the annual *Best American Sports Writing* and *Best American Crime Writing* anthologies. A former longtime resident of Oxford, Mississippi, he currently lives in New Jersey. His latest novel is *Want Not*.

Martha J. Miller is a writer and the proud owner of a dog that's a mix of pit bull loyalty and Chihuahua anxiety. She has written for *Oxford American*, the *Washington Post*, and *National Geographic Traveler*. Along with her husband and son, she happily calls Virginia home.

Ben McC. Moïse is a retired game warden who for twenty-four years patrolled the coast of South Carolina between the North Santee and South Edisto Rivers. A freelance writer and author, he penned a memoir, *Ramblings of a Lowcountry Game Warden*, and edited *A Southern Sportsman: The Hunting Memoirs of Henry Ed-*

wards Davis. He is a contributor to *Garden & Gun* and other publications. He lives in downtown Charleston with his wife, Anne, and their Boykin spaniel, Belle III.

LAWRENCE NAUMOFF teaches creative writing at UNC in Chapel Hill. He is the author of seven novels, including *Taller Women*, a *New York Times* Notable Book of the Year; *Silk Hope, NC*, which was made into a film starring Farrah Fawcett; and *Rootie Kazootie.*

T. EDWARD NICKENS reports on conservation, the outdoors, and Southern culture for some of the world's most respected publications. He is editor-at-large for *Field & Stream* and a contributing editor for *Audubon* magazine. In addition to writing for *Garden & Gun,* he is a judge for the magazine's Made in the South Awards. From the Nicaraguan rain forest to the shores of the Arctic Ocean, Nickens has authored more than a thousand bylined articles and won more than two dozen national writing awards.

GEOFFREY NORMAN is the author of twenty books and countless magazine articles, many of them related to the outdoors and the South. He has often written about his fondness for bird dogs, including the book *Riding with Jeb Stuart*, an account of a seventeen-year relationship with a hardheaded, big-going English pointer. He now hunts with a much more user-friendly English setter by the name of Woodrow.

P. J. O'ROURKE is the author of, most recently, *The Baby Boom: How It Got That Way . . . And It Wasn't My Fault . . . And I'll Never Do It Again.* He lives in rural New Hampshire with his family and three hunting dogs (one of which can actually hunt).

ROGER PINCKNEY lives and writes on Daufuskie Island, South Carolina: no bridge, no yoga, no yogurt, no traffic lights, no traffic at all, and all the fast food has fins, fur, or feathers. He is the author of eight books, both fiction and nonfiction, and innumerable magazine pieces.

JULIA REED is a contributing editor for *Garden & Gun* and writes the magazine's The High & the Low column. She is the author of *But Mama Always Put Vodka in Her Sangria!*; *Ham Biscuits, Hostess Gowns, and Other Southern Specialties*; *Queen of the Turtle Derby and Other Southern Phenomena*; and *The House on First Street: My New Orleans Story*.

DANIEL WALLACE is the J. Ross MacDonald Distinguished Professor of English at the University of North Carolina in Chapel Hill, where he directs the creative writing program. His most recent book is *The Kings and Queens of Roam*.

LOGAN WARD is a regular *Garden & Gun* contributor and the author of *See You in a Hundred Years*, a memoir about his family's move from New York City to the Shenandoah Valley to live the life of 1900-era dirt farmers.

ASHLEY WARLICK is the author of three novels: *The Distance from the Heart of Things*, *The Summer after June*, and *Seek the Living*. Her fiction and nonfiction have appeared in such publications as *Garden & Gun*, *Redbook*, *Oxford American*, and *McSweeney's*, and she is the editor of *edible Upcountry*, a magazine focused on local and sustainable foodways in upstate South Carolina. She is the recipient of the Houghton Mifflin Literary Fellowship, and a fellowship in literature from the National Endowment for the Arts. She teaches fiction in the MFA program at Queens University in Charlotte, North Carolina, and at the South Carolina Governor's School for the Arts and Humanities. She is at work on her fourth novel, *In Hunger*.

DONOVAN WEBSTER is the author of nine books and is a contributing editor at *Garden & Gun*. He has written for the *New Yorker*, *Vanity Fair*, *National Geographic*, *Smithsonian*, the *New York Times Magazine*, and others. An avid trout fisherman and reluctant screenwriter, he lives with his wife and children in the country outside Charlottesville, Virginia.

CURTIS WILKIE is the author of three books, most recently *The Fall of the House of Zeus*. He teaches journalism and serves as a fellow at the Overby Center for Southern Journalism and Politics at the University of Mississippi.

About GARDEN&GUN Magazine

Garden & Gun is a national magazine that covers the best of the South, including its sporting culture, food, music, art, and literature, and its people and their ideas. The magazine has won numerous awards for journalism, design, and overall excellence. *Garden & Gun* was launched in the spring of 2007 and is headquartered in Charleston, South Carolina.